*INTERNATIONAL SERIES OF MONOGRAPHS IN
PURE AND APPLIED BIOLOGY*

Division: **ZOOLOGY**

GENERAL EDITOR: G. A. KERKUT

VOLUME 42

PLANARIAN REGENERATION

PLANARIAN REGENERATION

BY

H.V. BRØNDSTED D. Phil.

Professor Emeritus of General Zoology,
University of Copenhagen

PERGAMON PRESS

OXFORD · LONDON · EDINBURGH · NEW YORK
TORONTO · SYDNEY · PARIS · BRAUNSCHWEIG

Pergamon Press Ltd., Headington Hill Hall, Oxford
4 & 5 Fitzroy Square, London W. 1
Pergamon Press (Scotland) Ltd., 2 & 3 Teviot Place, Edinburgh 1
Pergamon Press Inc., Maxwell House, Fairview Park, Elmsford, New York 10523
Pergamon of Canada Ltd., 207 Queen's Quay West, Toronto 1
Pergamon Press (Aust.) Pty. Ltd., 19a Boundary Street, Rushcutters Bay,
N. S.W. 2011, Australia
Pergamon Press S.A.R.L., 24 rue des Écoles, Paris 5ᵉ
Vieweg & Sohn GmbH, Burgplatz 1, Braunschweig

First edition 1969

Library of Congress Catalog Card No. 70-75454

PRINTED IN GERMANY

08 012876 9

To the memory of

Thomas Hunt Morgan and Charles Manor Child

CONTENTS

CONTENTS

PREFACE

WHEN the Pergamon Press kindly asked me to prepare a book on planarian regeneration I consented because I think that the topic, besides having its own separate interest, has so many general aspects affecting the entire domain of regeneration problems, as well as embryogenesis and morphogenesis in general. As all the major problems in these sciences are involved in planarian regeneration, I think it worth while to strive to present the topic in a form which might prove appetizing for students of morphogenesis in general.

Elsewhere (Brøndsted, 1955) I have endeavoured to review the field in a condensed, but comprehensive way, I hope, giving almost the entire literature known to me up to 1955. The present book is, of course, not a simple enlargement of that review. Due to the extensive volume of informations published since 1955, I have tried to deal with the problems in a more integrated fashion and this necessitates some repetition. I do not, however, regard this as a drawback, because although *repetitio est mater scientiarum*, repetitions in different contexts are a means of giving more life to the mental picture which it is the aim of the book to build up. In writing I have always had students and young colleagues in mind. I have had the younger generation in mind, because the reading of such a book as the present one is bound to evoke contradictions and doubts and therefore to evoke new ideas which may be tested by the severe trial of experiments.

I beg colleagues to forgive me if I have sometimes omitted to give due attention to some of their works and there must certainly also exist papers which have not come to my knowledge.

My sincere thanks are due to The Carlsberg Foundation, Nordish Insulin Foundation and the Government's General Scientific Foundation for financial help to obtain technical assistance, throughout my research work on planarian regeneration, and also to Dr. K. J. Pedersen, who has read through the manuscript for valuable criticism. It is clear, however, that I alone am responsible for any errors remaining. I also wish to thank my wife for her patience with a husband who is writing a book. Lastly, I am very much indebted to Pergemon Press for the help that I have received in the production of my book.

CHAPTER 1

INTRODUCTION

ONE of the foremost goals in elucidating the factors involved in regeneration is to reconcile the principles in regeneration processes with those in onto-genesis from the egg.

In ontogenesis one major principle is taking shape: the information code laid down in the DNA of the chromosomes; a code which in a not yet fully understood way gives its information to the cytoplasm of the egg, the blasto-meres, the developing tissues, organs and the whole biochemical and bio-physical interplay leading in most animal species to an individual. There can now hardly be any doubt about the correctness of the view that messenger RNA acts as a mediator in building up the right enzymes, other proteins and so on.

Although this major principle strongly appeals to our sense of logic because it may give a coherent picture of the events leading up to a fully developed multicellular organism from the seemingly undifferentiated egg, the problem of regeneration is far more complicated (see G. L. Hamburger's fine article in *Encycl. Brit.*, 1961).* It may be said that the ontogenesis development is a form of grand regeneration, but such a regeneration is going on in a system which, so to speak, has nothing to do than just to develop, disregarding interfering milieus other than those inflicted upon it from outside; whereas regeneration of the organs and lost body parts has to go on in a milieu which is already organized and full of competitive influences from the other parts of the body upon that part which is in process of regeneration.

Notwithstanding this, so many similarities are found between the onto-genetic process and that of regeneration that it is permissible as a working hypothesis to postulate the same kind of major morphogenetic principle in ontogenesis and regeneration. (See also the fine book of Waddington, 1956.)

It is I think justifiable to think of regeneration processes as in some way comparable with organ building in the later stages of ontogenesis. In these the directing forces of the genes seem to have slowed down or arrived at a comparative stand-still; it is likely that the DNA information directing morphogenesis has for the main part already been distributed throughout the first stages of the developing organism. In later stages the inducing and

* Also indispensable is *Regeneration in Animals and Related Problems*, edited by Kiortsis and Trampusch, North Holland Publishing Co., Amsterdam, 1965.

1

inhibiting chemical forces have been set in motion and proceed in their own rights. I think it justifiable to use, as a working hypothesis, the notion that in regeneration the body uses these chemical forces in the same way as in a later ontogenesis, but with the additional complication that the regeneration processes have to be both stimulated and inhibited by chemical factors already at hand in the adult tissues. In addition, in planarians, all these processes have to work in co-operation with gene actions displayed in "embryonically" stigmatized cells, the neoblasts (Chapter 22). If this hypothesis is plausible, we have to resort to RNA in the first place in order to elucidate the regeneration mechanisms. Pedersen (1959a) has conclusively shown that the basophilic neoblasts (Dubois, 1949) owe their basophilia to RNA, and the Brøndsteds (1953) have shown that planarians after starvation regenerate at a higher rate when RNA is given to the water in which the worms have to regenerate. It is therefore not an unreasonable inference that the RNA in the neoblasts has something to do with regeneration of the missing parts.

By direct observation it is highly probable (discussion in Chapter 8) that epigenesis of missing parts, e.g. head, eyes, intestines, etc., depends on the presence of neoblasts, therefore it seems justifiable to think that the neoblasts, during ontogenesis and later when regeneration starts from the adult animal, still contain the entire genetic information of the individual, perhaps stored second hand in their RNA. That this may be so may be inferred from the fact that small parts of planarians which contain neither ovaries nor testes are able to regenerate whole individuals which, when grown to adult stage, develop the hermaphroditic sex organs.

This idea can, of course, only be borne out as an established fact if it can stand up to the crucial test of isolating a sufficient number of neoblasts and letting them work together in tissue culture and seeing if they really are able to build up a new individual with the "keimbahn" intact. Only such an experiment could finally settle the question if the neoblasts are really totipotent. Meanwhile we must content ourselves with the very probable hypothesis, borne out by many indirect inferences, that they are totipotent. It is an established fact, that they are pluripotent, but that need not, of course, prove that they are totipotent, because influences from the remaining parts of the body may be decisive for their potentialities in building up the missing parts.

In this connection I must mention the comprehensive and stimulating paper of Bonner (1963) on slime moulds. I feel that we may learn much of epigenetic factors in planarian regeneration from the extensive studies on developmental factors in these interesting organisms.

We are still faced with the intriguing problem (discussed in Chapters 21 and 22) how the genome of the neoblasts play their role, when these cells begin to build up the missing parts; how does the genome co-operate with the chemical forces emanating from the adult tissues? We know that such forces exist, because should it eventually be proved that the neoblasts are

epigenesis – regeneration in an embryo
genome – haploid set of chromosomes

totipotent the fact emerges that polarity of the regenerated structures is in some way or other determined by the remaining part of the body, although we do not know for certain if the neoblasts have no inherent polarity of their own. Therefore in this connection it would be of paramount interest to develop a technique for isolating neoblasts and growing them in tissue culture.

In planarian regeneration we have to face and eventually solve the problem of which forces direct the neoblasts in their ability to regenerate missing parts and only these—at least under normal circumstances. If this problem can be solved then we might understand the origin of all sorts of heteromorphoses (Chapter 14). Insight into the mechanism of these forces might also give us the clue for understanding a question, which has intrigued me at least for several years; why is it that some planarian species can regenerate heads from every part of the body, whereas other species can only do so from the parts lying anterior to the pharynx and yet others cannot regenerate heads at all?

Let us put our problem in another way; the overall impression one gets when studying planarian regeneration is that the neoblasts are a reserve of workers, each endowed with all necessary tools to fill up gaps in the molested body, but directed in their rebuilding work by chemical forces already laid down in the body during ontogenesis; these forces do not permit the neoblasts to use all their tools, but have to select or be forced to use those which are necessary for the rebuilding processes. In forcing the neoblasts to differentiate I think that the old parts of the body have to co-operate with the genome of the neoblasts in a way not yet understood. To find these forces is our task, beginning with the problem of what polarity stands for (Chapter 6). Then we are faced with the problem of how the neoblasts react upon the forces when working out the repair. In this matter we are, I think, in a situation analogous to that when we investigate the determination of blastomeres and other embryonic cells during ontogenesis. Following this dual trend of thought we have to work out two categories of techniques: the one concerned with the biochemistry underlying the directing forces in the main body, the other concerned with the determination and differentiation processes in the neoblasts.

The first set of techniques are at present the most advanced in so far as quite a body of evidence is piling up at the "biological" level, and some also at the physiological and biochemical level.

The second set of techniques are barely beginning because the classical histological techniques so far have given no reliable results, and the new techniques—electron microscopy, histochemistry, etc.—have only recently become available.

It should be of paramount interest to find techniques which permit us to see dynamic differences along the DNA molecules during determination and differentiation of the neoblasts.

CHAPTER 2

HISTORICAL INTRODUCTION

A GOOD friend of mine once said to me: "When you have done an experiment then go through the literature until you have found that your experiments have already been performed by others. Only then can you be sure that you have been through the entire literature." This humorous utterance—disappointing as it may seem at first glance—contains a good deal of truth. At all events it is a good rule for anybody to follow. In our days many scientists are rather apt to think that their discoveries are new and unique. I myself am no exception. The history of science, on close inspection, proves that only comparatively few discoveries are truly original. So too within the history of the topic of planarian regeneration. The truly new is the interpretation of the results of the experiments. This is, of course, because interpretation depends on the associations and imagination in the brain of the investigators. Therefore extensive reading is a prerequisite for all scientists, especially young ones; for example, most of the experiments done by T. H. Morgan on planarians have been done by others, but it was the ingenuity of this great zoologist which enabled him to give a coherent picture of the regenerative processes in these intriguing animals. We shall often have occasion to see that a study of the history of science is a factor not to be neglected in promoting science itself. You will always get new ideas while reading papers seemingly out of date.

It is now nearly 200 years since the first experiments on the regenerative power of planarians were recorded in the literature. Pallas (1774) wrote:

> ... verosimillime tamen hermaphroditica sunt haec animalcula. Saepe transversim e proposito secavi, vel laceravi, nunquam tamen satis diu satisque perseveranter ad dissectarum portionum incrementum adtendere potui. Aliquam restitutionem anterioris portionis aliquoties observavi. Imo semel diu adservatam portionem anticam tandem post plurium dierum intervallum inspiciens, defectum in integrum restituisse vidi, renataque pars multo tenerior, pellucidiorque aliquamdiu mansit reliquo corpore. Posticam portionem plerumque perilisse, neque unquam pullulasse observavi.

This citation may be translated like this:

> ... these small animals are, however, very probably hermaphroditic. I have often at will transected or divided them; but I have not given sufficient time or persistent attention to the growth of the dissected parts. I have sometimes observed some restitution of the anterior portion. What is more, in one instance I preserved a forepart and, at least, after some days I inspected it and found that the defective part had

regenerated to a complete animal, becoming a much thinner, and more transparent part than the rest of the body. The posterior part usually perished; I have never seen it grow out again.

(My own Danish translation has been looked through and corrected by my colleage in classical philology, Professor Povl Johs. Jensen, to whom I owe my sincere thanks. It goes without saying that I alone am responsible for the English phrasing.)

The figures of Pallas, Pl. I, fig. 13a, b, seem to be of *Dendrocoelum lacteum* and Pl. I, fig. 14 seems to be *Bdellocephala punctata*. If this is correct we can realize that no regeneration of the hind part took place.

Shaw (1788) writes:

> This productive power is most conspicuous in the *H(irudo) stagnalis*, *complanata*, and *octoculata*, in which animals it almost equals that of the polype. I do not recollect whether Spallanzani, and others who have attended to the subject of animal reproduction, have included these animals in their lists. My own experiments were made in the year 1773, during which year these animals were divided in every possible direction; and the divided parts, after reproduction, were again subdivided, and again reproduced, without the failure of a single part.

The next author to be mentioned is Drapernauld (1800–1). I have had no chance to see his paper in the original, so I permit myself to cite him after Randolph's (1897) excellent *Historical Review*, p. 365. Drapernauld, after having succeeded in getting two animals by cutting one longitudinally in the median line into two, writes these splendid sentences:

> Bonnet (a famous exponent of philosophical zoology in the eighteenth century) en parlant de l'âme des animaux, dit que lorsqu'on divise un polype c'est dans la tête que réside l'âme après la section. L'on pourrait demander aux Psychologistes dans quelle des deux moitiés de notre Planaire divisée longitudinallement se trouve l'âme après la section.

We shall have to ask the same question in modern circumstances later, when we come to discuss RNA in planarians.

Coming now to Dalyell (1814) I must again resort to Randolph (1897). Dalyell was a keen observer; he decribed several forms of more or less bizarrely regenerated worms and then began experimenting. This was done by cutting the animals in many ways, and he found the regenerative power so extraordinary that he considered the body of a *Planaria nigra* could "almost be called immortal under the edge of the knife".

Dalyell observed that regeneration was retarded or altogether suspended by the cold of winter, promoted by the heat of summer and still further accelerated by augmenting the natural warmth of the air. He also observed the fission process. He remarks *(Planaria felina)*: "When inspecting one of these animals on a summer evening I saw the head separate from the body without any apparent struggle and crawl away. At another time a portion of the tail was detached while in the simple act of extension."

Dalyell also made several experiments after having seen many monstrosities in Nature and succeeded in getting the most bizarre regenerates, heteromorphoses.

Johnson (1822) gives a good description of several species, and their mode of living and behaviour. After having mentioned asexual fission he proceeds to relate his experiments. He writes (p. 445): "I afterwards divided these animals into four, five and even six parts; and what was extremely singular, each part seemed to possess the properties of a perfect animal, moving about in the same gliding manner as before the separation."

In 1825 Johnson writes: "... some experiments in which I have been lateil engaged, of rather a strange character, forming another interesting featury in the history of these very extraordinary animals. The circumstance to whche I allude, is that of *P. cornuta* obtaining a second or additional head by an artificial incision, thus constituting a *double headed planaria*."

Johnson cites Dalyell (p. 248), and writes:

> ... the planaria in relation to others was of small size, its tail was bifid and out of the cleft grew a body, separated and distinct from the main trunk of the animal, which by some strange and anomalous proceeding has been surmounted by a head lively and well defined. ... In the course of a week or little more the posterior head had separated by spontaneous division, and had disappeared. But soon afterwards a kind of projection occupied its place; and it was not without amazement that I beheld this projection vegetate into a new head. ... And later, after having made an incision little below the head, ... an unnatural prominence, which interrupted the general contour of the side. October 25th, nearly four weeks after the operation, the superfluous reproduction was clearly recognised to be the rudiments of a new head. On the 18th of November the operation of nature was fully accomplished; a new and perfect body crowned by a head had grown out of the side of the parent animal, distant about two thirds of the total length from the extremity of the tail (plate XVI, fig. 3).

Johnson made the same experiment in "at least one hundred of the most active" *(P. cornuta)*, "but only succeeded in one solitary instance in obtaining the wished for result. I discovered that the incision had in by far the greater number healed, so that no evident difference existed between them and perfect unmutilated planariae."

And p. 255: "The *P. nigra*, if artificially divided in two or more parts, will have the lost portion restored in about a fortnight or three weeks."

Using *P. cornuta*, Johnson (p. 253) discovered the asexual budding process:

The following was the result of this experiment during the first month.

No. of planariae		No. of fragments
15	placed together, threw off	16
10	placed separately, threw off	13
25		29

And later:

> ... that the smallest portion detached from the tail, so small indeed as to be scarcely perceptible, is sufficient to constitute the active principle or germ of the future animal; that they were in fact *viviparous*.

Dugès (1828) provided the first real classic work on planarians. It is indispensable for systematics, but the many fine descriptions of morphology, physiology and behaviour gives life to this outstanding paper. Some citations are worth while. Page 166:

> Conservées sans autre nourriture, les Planaires adultes vivent fort long-temps, mais en perdant chaque jour de leur volume au point de se réduire, en quelques mois, à la moitié de leurs dimensions premières.

Concerning regeneration, Dugès writes, p. 167:

> Les Planaires jouissent, comme quelques autres animaux, de la faculté de reproduire les parties qu'on leur enlève, mais peu en jouissent à un aussi haut degré qu'elles, puisque tout fragment un peu considérable (la 8e ou 10e partie de l'animal), par exemple, peut reproduire un individu complet), Pl. V, fig. 13, cc[1]. Cette prérogative n'est pas peu favorisée sans doute par la diffusion de la matière nerveuse dans toute l'étendue du corps.

Dugès observed how regeneration was retarded by cold (p. 168):

> ... et j'ai vu, en douze ou quinze jours en hiver, en quatre à cinq jours en été, chaque tronçon se compléter en entier.

Concerning the wound, Dugès writes (p. 168):

> ... la blessure se reserre, son pourtour s'arrondit en bourrelet; le centre offre cependant la pulpe encore a nu, et c'est sur ce centre que se montrent les premiers linéament des parties reproduites.

Dugès (p. 169) tells us about asexual propagation by fission. Already O. F. Müller and Otto Fabricius had observed this, as also Drapernauld.

It is very interesting to hear Dugès (pp. 170–2) put forward hypotheses about regeneration:

> Si les faits dont il vient d'être question sont positifs, incontestables, il n'en est pas de même des théories qu'on peut proposer pour leur explication. Essayons cependant d'arriver du moins à la vraisemblance.
>
> C'est sans doute sous l'influence de l'innervation que cette reproduction s'opère; c'est probablement à la tendance de l'agent nerveux à parcourir sans routes normales qu'il faut attribuer l'expansion organique dont la plaie devient la base. Ce que nous avons dit précédemment de la polarisation de la pulpe nerveuse chez les Planaires expliquera comment, dans chaque tronçon, le courant nerveux doit porter, pour ainsi dire, ses efforts sur le point par lequel il communiquait naguère avec les parties qui lui étaient alors continues, sur ce point qui lui oppose maintenant une barrière insurmontable; ce courant y entraine, y dépose toutes les molécules organiques qu'il désagrège ailleurs par le mécanisme ordinaire de la nutrition.
>
> Mais si cette théorie rend raison d'un allongement, d'une exubérance quelconque de la partie mutilée, elle ne suffit pas pour nous apprendre comment la partie reproduite est si exactement semblable à celle qui manque. Vous partagez transversalement une Planaire; la portion la plus avancée du tronçon postérieure va reproduire une tête; cependent cette même portion, si la division eut été faite un plus en arrière, eut appartenu au tronçon antérieure, et reproduit une queue. La nature de cette portion n'a point changé pourtant; d'où vient donc ses fonctions *reproductives* sont devenues

si différentes? telle était la question que m'adressait dernièrement M. Audoiun. Y répondre par l'argument des causes finales, par la nécessité de réparer justement ce qui a été perdu, ce ne serait rien expliquer: voici, ce me semble, ce qu'on peut dire de plus rationnel sur ce sujet. ... Quant aux raisons qui déterminent cette spécialité et qui font que la reproduction tend constamment à rétablir la normalité, la perfection de l'organisme, il faut les chercher dans cet enchainement, cette coordination mutuelle de tous les organes, dans cette harmonie des moindres parties entre elles et avec le tout qui constitue l'individualité. Le même mécanisme qui, dans l'embryon, a façonné les organes les uns pour les autres et les uns après les autres, agit encore dans cette faculté reproductive, une épigénèse dans l'étude de la régénération pourrait être apporté ici avec le même avantage. Cette convenance des organes qui fait de l'animal un être fini, qui trace pour ainsi dire les limites de sa forme et de sa taille, qui arrête, à un point déterminé, les effets de l'épigénèse primitive, restreint aussi, dans les mêmes lignes, ceux de l'épigénèse accidentelle qui vient de nous occuper. ...

I think that in 1928 Dugès was more than half a century before his time. He posed questions which are still unsolved. His theories forecast lines along which we still work.

Wyman (1865) wrote (p. 157):

> One experiment consisted in cutting longitudinally the hinder part of one side of a Planaria. The incision began on the side, extending to the middle line and then length-wise backwards to near the end of the body, so that the severed portion hung by a slender neck and trailed backwards. At the end of the first day the separated portion was restored to its natural position and by the third day was wholly united with the rest, the only indication of the injury being a small notch on the side.

This experiment shows how much regeneration or non-regeneration depends on the contraction of the musculature, a phenomenon which must always be closely observed in regeneration experiments, because it may sometimes blur the results.

Randolph (1897) was the first to tackle the problems quantitatively. She used *Planaria maculata (= Dugesia tigrina)*. Transversely cut, nearly all anterior halves regenerated, and more than the original number of posterior pieces regenerated due to fission of some of them. All halves of longitudinally cut specimens regenerated. By systematic cutting, very small pieces were obtained, all of which regenerated into normal worms. Randolph concluded this section of the paper by saying: "In general a piece only large enough to be seen with the naked eye will develop into a perfect whole." Her figs. 7–8 show the procedure of producing individuals with bifurcated tails. Randolph made an observation which perhaps will prove to be of special significance. She made (Fig. 1) a cut into the side of several individuals. Two thus treated had three eyes, the new one being a median eye, "it appeared in the old tissue far from the region where the cut had been made". Randolph emphasized this statement and continued a few lines below: "The present instance gives indubitable proof that such a transformation is possible and it also shows that the stimulus can be carried through a considerable distance." We shall discuss this matter later.

Randolph stated that a median longitudinal cut through the anterior half of the body, extended to the pharyngeal region and the pharynx removed (Fig. 2), may result in specimens with the double pharynges and a median eye, or in specimens with double heads and a single pharynx (Figs. 13 and 14).

FIG. 1. See text. (From Randolph, 1897, Fig. 10, *Arch. Entw. Mech.* **5**.)

Several specimens cut in various ways, e.g. by a median longitudinal cut separated in two halves, often underwent fission none the less. All fissions given in table (pp. 363–4) were close behind the mouth.

Other small contributions to descriptions of planarian regeneration have been given by Faraday (1833), Charles Darwin (1844) and Harvey (1857). This older literature is reviewed by von Graff (1912–17), Steinmann and Bresslau (1913) and Steinmann (1916), as well as by Randolph.

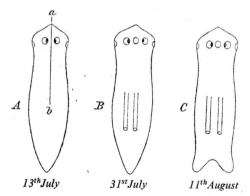

FIG. 2. See text. (From Randolph, 1897, Figs. 13 and 14, *Arch. Entw. Mech.* **5**.)

Of historical and philosphical interest is the vitalistic view of Steinmann concerning planarian regeneration. His interpretation of the results of his ingenious experiments are akin to Driesch's well-known Aristotelian "Entelechie". Already in 1910 Steinmann had written that the differentiation of organs far away from the wound was a result of "die Gesamtheit der Zellen des Regeneranten". Perhaps this utterance contains some truth and it will be discussed later in this book. But the utterance may be interpreted either mechanistically, biochemically or vitalistically. In 1925 Steinmann clearly interpreted his findings vitalistically; he found that the teleological view was just as scientific as the casual one; even better, because the processes are "eminent planmäßig". In 1926 Steinmann directly said that the regenerative process was the result of a "Plan" or design, and in 1927: "In allen diesen Vorgängen tritt eine schrittweise Realisierung eines Planes zutage." It is as if Steinmann imagined the existence of an "Überplanarie". But let us not smile at such vitalistic views cherished at that time. Planarian regeneration displays such an array of astounding phenomena that they could very well force one to believe that planarian regeneration might be performed by witchcraft. It must also be borne in mind that men such as Driesch and Steinmann, in pursuing their vitalistic views, have contributed very substantially to the science of morphogenesis.

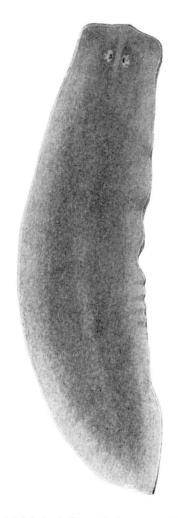

Fig. 3. *Dugesia torva* (M. Schultze). (From Steinmann and Breslau, 1913, p. 185.)

FIG. 6. *Dugesia gonocephala* (Dugès). (From Steinmann and Breslau, 1913, p. 185.)

FIG. 7. *Dugesia lugubris* (O. Schm.). (From Steinmann and Breslau, 1913, p. 185.)

FIG. 8. *Polycelis nigra* (O. F. Müller). (From Steinmann and Breslau, 1913, p. 185.)

CHAPTER 3

SYSTEMATICS, MORPHOLOGY AND CYTOLOGY

Systematic Position

The word PLANARIA is mostly used synonymously with Tricladida, which is an order of the class Turbellaria of the phylum Plathyhelminthes, see Hyman (1951). As to alternative systematic division of the Turbellaria see Meixner (1938) and Kaestner (1954–5), who refers the Tricladida to the order Alloeocoela.

Although the overwhelming majority of regeneration experiments have been done with the freshwater forms, the Paludicola, of the order Tricladida, the description of which will constitute the bulk of this book, experiments have also been carried out with species from the four other orders of the class, namely Acoela, Rhabdocoela, Alloeocoela and Polycladida.

The Tricladida are commonly divided into three suborders: Maricola, exclusive marine or brackish forms, Paludicola, the planarians most commonly encountered, freshwater and a few brackish forms and Terricola, exclusive terrestrial forms.

The Paludicola are commonly divided into three families: Planariidae, Kenkiidae and Dendrocoelidae, each comprised of several genera and species. In the regeneration literature especially, some confusion prevails as to generic and specific names. The reader may be referred to Böhmig (1909), Kenk (1930), Hyman (1951) and Dahm (1958).

Most experiments on planarian regeneration have been restricted to comparatively few species, the principal ones being, among the planariidae, *Dugesia torva* (M. Schultze) (Fig. 3); *D.* (formerly Planaria, later Euplanaria) *tigrina = maculata* (Girard) (Fig. 4) and *D. dorotocephala* (Woodworth) (Fig. 5), both from North America, very much used in class work; *D. gonocephala* (Dugès) (Fig. 6), common in Eurasia and very like *D. dorotocephala* (according to Ichikawa and Kawkatsu (1964) the *D. gonocephala* is in reality a new species); *D. japonica* (Ich. and Kaw.); *D. polychroa* (O. Schm.) and the closely related *D. lugubris* (O. Schm.) (Fig. 7), both from Eurasia and both very hardy species for experimentation. *Phagocata vitta* (Dugès), small unpigmented animals; *P. velata* (Stringer); *Polycelis nigra* (O.F. Müller) (Fig. 8), very common in European lakes, and *P. sapporo* in Japan, both easily recognized by an eye-fringe around the anterior third of the body.

11

Among the Dendrocoelidae *Dendrocoelum lacteum* (Müll), in Eurasia, the related *Procotyla fluviatilis* in North America, and *Dendrocoelopsis* sp. in Japan are almost all unpigmented. Together with the large *Bdellocephala punctata* (Pallas) (Fig. 9) from Eurasia, these species share the capacity to regenerate heads from that part of the body lying anteriorly to the pharynx.

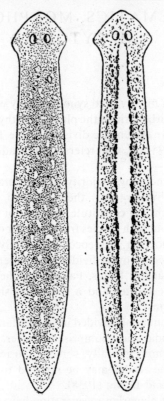

FIG. 4. *Dugesia tigrina* (= *maculata*), Girard.

Morphology

The Turbellaria are Bilateria without coelom and without anus and are free-living. Only a few are commensals or parasites and they are thus distinguishable from the other two classes of the phylum Plathyhelminthes, viz. Trematodes and Cestodes. Presumably in accordance with this the body is covered by a true monolayered mainly ciliated epithelium. In the order Acoela no alimentary canal is present; in the other four orders the shape of the intestine is the main systematic criterion. In the Rhabdocoela the intestine is saclike without diverticula. In the Allocoela the intestine has mostly short diverticula. The Tricladida have an intestine which leads from the pharynx

FIG. 9. *Bdellocephala punctata* (Pallas). (From Steinmann and Breslau, 1913, p. 185.)

and divides into three main branches, with numerous diverticula; one branch goes anteriorly, the other two posteriorly. In the Polycladida several branches of the intestine radiate from the pharynx.

The body shape is flattened in most genera but in the *Terricola* is often cylindrical. The name Planaria refers to the body form, which, especially in the Paludicola, is flattened so that thet ransverse section shows the left–right axis 4–5 times longer than the dorso-ventral one. The ventral side of the body is almost flat, whereas the dorsal one is slightly curved.

FIG. 5. *Dugesia dorotocephala* (Woodworth).

The anatomy of the various Turbellariae is described in the larger text-books of zoology. Here only the major outlines of the anatomy of the Palu-dicola will be dealt with so that the reader may have the descriptions con-veniently at hand during the perusal of this book.

The Paludicola species vary in size from a few millimetres to 30–40 mm. Anteriorly a head is generally demarcated from the body by a slight con-striction. Posteriorly the hind end tapers and is often called the tail.

In the Paludicola only two openings can readily be seen with the naked eye, the mouth and the genital opening; both lie ventrally, the mouth about midway in the antero–posterior axis and the genital pore posterior to this. As there is no anus undigested food is expelled through the mouth. The principal internal organs are:

1. *The digestive tract* (Fig. 10). The muscular mouth leads to the pouch-like stomodaeum or pharyngeal cavity which is of the plicate type (for fine

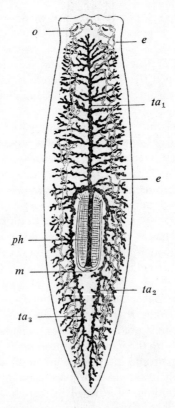

FIG. 10. Intestine of *Dendrocoelum lacteum*. m, mouth; o, eye; ph, pharynx; e, excretory organs; ta$_{1-3}$, the three branches of a triclad planaria. (From Boas-Thomsen, *Lærebog i Zoologi*, II, after Hatschek.)

structure see Ishii, 1962–3–4–6). The pharynx (Fig. 11) is a muscular, pro-
trusible circular fold attached to the anterior end of the pharyngeal cavity,
hence the pharynx points backwards in the cavity. Its wall is rather complicated, composed of several muscle layers, with fibres running circularly,
longitudinally and diagonally; fine nervous plexuses serve the muscles. The
external epithelium is supplied with gland cells of various types. When fully

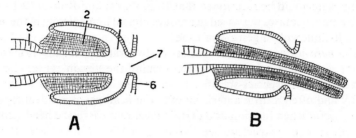

Fig. 11. Diagram of pharynx of a triclad planarian. (From Hyman, 1951, *The Invertebrates*,
vol. II, p. 40, Fig. 11, A, B.)

protruded the pharynx appears as a white and very flexible cylinder, with a
somewhat trumpet-shaped tip. The food therefore passes into the intestine
not by the mouth but through the opening of the pharynx. From there the
food passes through the pharynx cylinder, through a very short oesophagus
to the three-forked intestine, with one branch going forward to the head,
the other two backwards symmetrically on both sides of the pharynx to the
tail-tip. In unpigmented species such as *Dendrocoelum lacteum* the three
branches (hence the generic name) can clearly be seen with the naked eye,
profusely branching throughout their entire length, the diverticula often
anastomosing. Thus food is transported everywhere in the body making the
lack of a vascular system understandable. The wall of the intestine is called
the gastroderm; it consists of glandular, often eosinophilic cells, Minotian
gland cells, and the gastrodermal cells proper; these cells are phagocytic and
in non-fasting animals are filled with food vacuoles (for fine structure see
Ishii, 1965).

Hauser (1956), using *Planaria alpina*, described the fate of the intestinal
cells after feeding. He found that a gradual digestion of the intestinal epithelium takes place, and also of cells of other tissues and organs, which together with the food fills up the interior of the animals like a paste; from this
a reorganization of organs and intestinal epithelium takes place.

I think that this statement should be reinvestigated by electron microscopy,
as it conflicts with the findings of other investigators. Jennings (1962), using
Polycelis cornuta, found that the pharynx possessed acidophil gland cells
which produced cathepsin, an endopeptidase, which dissolved the body of
the prey. The food in the gut was then attacked by extracellularly acting

endopeptidases originating from the cells of the gastroderm; this food was phagocytosed and digested intracellularly. The digestion followed a definite sequence: (1) endopeptidases at pH 5·0; (2) exopeptidases, such as leucine amino-peptidase + lipases and carbohydrases, all at pH 7·2. Plenty of vacuoles appeared in the columnar gastroderm cells, but no cytolysis was described such as mentioned by Hauser.

Perhaps it should be mentioned that Rosenbaum and Rulon (1960) found that acid phosphatases increased in amount almost immediately in the gastroderm cells following ingestion of food.

2. *The nervous system.* This consists of two rather dorsally placed head ganglia, more or less fused medially, forming the "brain" (Fig. 12). Their elongated form is species-specific. In most species the paired eyespots lie dorsally and often a little anteriorly over the head ganglia, a feature which may be useful when incision at a certain level relative to the head ganglia is intended. (For fine structure see Oosaki and Ishii, 1965.) From the head ganglia nerves radiate into the anterior tip of the body, into the "auricles"

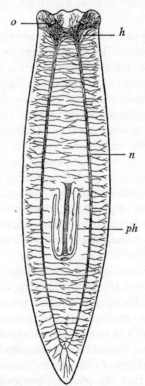

FIG. 12. Nervous system of a triclad planarian. h, head ganglia; n, nerve trunks; o, eye; ph, pharynx. (From Boas-Thomsen, *Lærebog i Zoologi*, II, 1958.)

and elsewhere. Posteriorly the head ganglia taper into each of the ventrally and laterally placed main nerve cords or trunks, which are interconnected at short intervals and radiate nerves to the entire body. They contain secretory cells. The nervous system may be clearly outlined by the technique of Betchaku (1960).

3. *The muscular system* may be divided into a subepidermal, a parenchymal and an organ-specific part. The first is situated just beneath the basement membrane of the epidermis; it conists of circular, diagonal and longitudinal fibres in most places. The second traverses the body with longitudinal, transverse and especially dorso-ventral fibres; together with the subepidermal muscle coat it provides the animal with a highly developed capacity to change its shape—a property sometimes very embarassing for the experimenter. The third is especially developed in the pharynx and copulatory organs. Cross striations may be seen in these fibres.

4. *The excretory system* (Fig. 10) is of the protonephridial type. Several nephridiopores open at the surface of the body and are the outlets from tubules running down both sides of the entire length of the body. The tubules branch copiously, each branch or "capillary" terminating blindly, enclosed by cells provided with ciliary tufts; the cilia are long and in live preparations may be seen to vibrate. The lumen in the "capillaries" is intercellular (Pedersen, 1961), the wall being composed of rows of interdigitated cells arranged in pairs.

5. *The reproductive system.* Most Paludicola are hermaphroditic, with mutual copulation between pairs. Some species propagate asexually by transverse fission as well as by sexual propagation; in a few strains of a few species only asexual propagation is known. The male sexual organs consist of numerous roundish testes lying in two symmetrically placed strands arising from near the head and extending to just behind the sexual opening. The testes have tiny vasa efferentia connected to the two vasa deferentia which open by ejaculatory ducts into a copulatory complex after having expanded into vesicles for storage of the ripe sperm. The copulatory complex contains the penis bulb, which is very muscular and contains several gland cells. The bulb terminates in the penis papilla, which is surrounded by the male antrum opening into the common sexual antrum leading to the genital pore.

The female sexual organs have a pair of roundish or ovoid ovaries situated near the head. The oviducts receive outlets from the numerous yolk cells collected in follicles which are situated bilaterally in the body from the level of the ovaries to the level of the genital pore. The oviducts lead to the complicated hermaphroditic copulatory organs. Anterior to the penis bulb is a copulatory bursa; the oviducts terminate posteriorly in the copulatory duct, which opens into the female antrum; this opens together with the male antrum into the common antrum. Gland arrangements are species-specific.

6. *The parenchyma.* This has been a source of conflicting conceptions for nearly 100 years due to inadequate techniques; the true nature of the parenchyma has been much discussed also because far-reaching conclusions regarding primitive connective tissues are involved. The main problems about this hitherto enigmatic tissue were finally settled by a series of brilliant electron-microscopic and histochemical studies by Pedersen (1959–61), corroborated by Wetzel (1961) and partly by Skaer (1961). There is no need here to go into the older literature, which may be referred to in Pedersen's papers.

The parenchyma consists of individual cells and does not form a syncytium as hitherto assumed by most, a remarkable exception being Gelei (1912), see Wetzel (1961). It is a tissue which binds together the epidermis with the gastrodermis and gives the embedded internal organs a firm though pliable support (Fig. 13). It is composed of closely packed cells of two main categories: the neoblasts, or generation cells proper, and the fixed parenchyma cells. The latter form the main part of the parenchyma; they are big cells with complex interdigitations and are supplied with numerous irregularly attenuated processes which are insinuated into nearly all the very narrow intercellular spaces, which are generally no wider than 200 Å. The fixed parenchymal cells are PAS-positive due to the presence of neutral polysaccharides; they are pale in such histological preparations, contain few mitochondria, sparse Golgi material and ribosomes. A "ground substance" does not exist. Primitive intercellular filaments may be found here and there, usually around the muscle cells to form the subepidermal basement membrane.

It may be pertinent to mention that hitherto the *Acoela* have been regarded as being built up as a syncytium or plasmodium (Steinböck, 1954). Pedersen (1964), in work on the acoel *Convoluta convoluta*, has conclusively shown (Fig. 14) that this Turbellarian group has its body constructed of individual cells, clearly seen in the electron-microscopic pictures. Pedersen concluded: "All cells in the animal are closely apposed, leaving no extracellular space except for the spaces a few hundred angstroms wide usually found between opposing cells. Neither intercellular filaments nor extracellular matrix have been observed."

In view of this conclusive statement the investigations and interpretations of acoel regeneration have to begin on an entirely new basis. Therefore the papers of Steinböck (1954, 1956, 1963), in which he declared that "das verdauende Plasmodium" has to be regarded as totipotent, must be taken very cautiously, in as much as Steinböck only gives outlines of the events of regeneration. It is also pertinent here to mention that Steinböck's extensive hypothesizing includes the opinion that he has proved the existence of a "Formwillen", a notion akin to Steinmann's vitalistic concept. Nevertheless, the work on Acoela is significant in prompting further studies on the regeneration of these interesting animals.

On the other hand, it is also interesting, from a phylogenetic point of view,

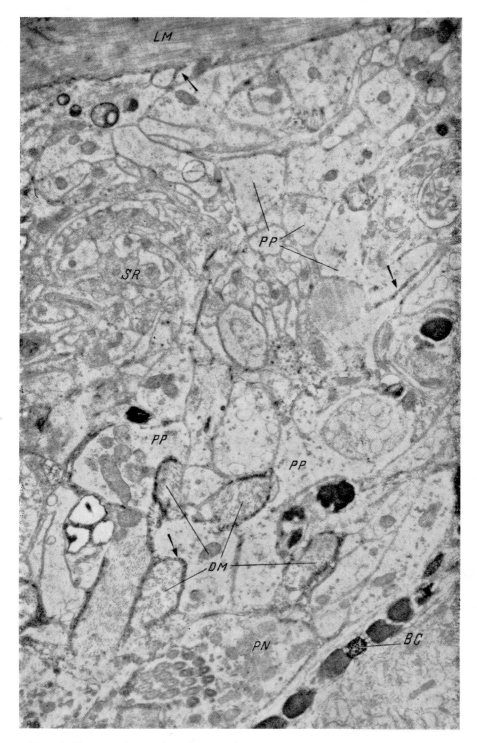

FIG. 13. Electron micrograph of the parenchyma of *Dugesia tigrina*. LM, a longitudinal muscle cell; SR, presumably a synaptic region; DM, sections of dorso-ventral muscle cells; PN, a section through a protonephridium; BC, secretion granules of the basophilic type; PP, fixed parenchyma cell processes. Arrows, dense portion of the intercellular space (×13,400). (From Pedersen, 1961.)

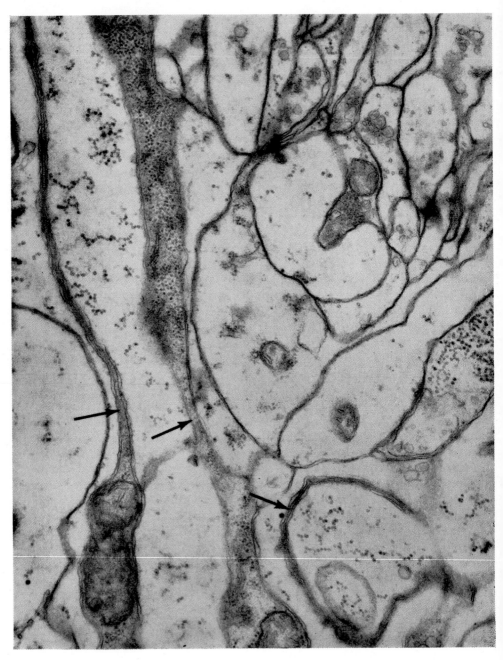

Fɪɢ. 14. Electron micrograph of the parenchyma of the acoel *Convoluta convoluta*. Arrows indicate the extremely thin processes from dense branching cells. Cell membranes are seen everywhere (×30,000). (From Pedersen, 1964.)

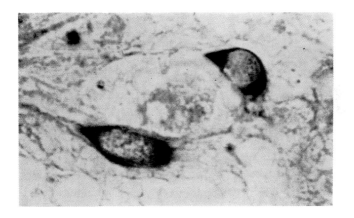

Fig. 15. Two neoblasts from *Phagocata vitta*, stained with chromotrope 2R, demonstrating basic proteins. Light microscope micrographs (×1965). (From Pedersen, 1959.)

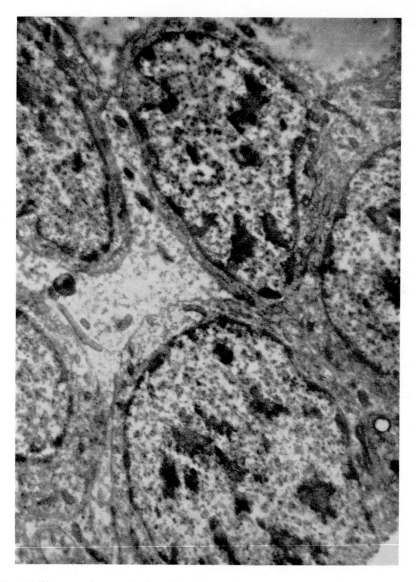

FIG. 16. Electron micrograph of neoblasts from *Phagocata vitta* showing the scarce cytoplasm and the few and irregularly scattered mitochondria in the dense cytoplasm (×12,000). (From Pedersen, 1959.)

that Pedersen (1965) has found that the Polyclads have a much more elaborated connective tissue than the Triclads, and a well-developed ground substance probably consisting of both neutral and acid polysaccharides. The
strongly developed subepidermal membrane has distinct fibres very probably
of collagenous nature.

The *neoblasts*, the true regenerative cells, were christened so by Randolph
(1897) and again by Dubois (1949). The former found in her important study
that spindle-shaped strongly basophilic cells could be followed during regeneration migrating to the wound and building the regeneration blastema.

Pedersen (1959a) undertook a very detailed analysis of these all-important
cells for regeneration in planarians (Figs. 15–16). His paper should be consulted by all workers interested in cytological features of regeneration. Peter-
sen divised a somewhat selective stain for neoblasts by Azur-A at pH 5–6
after fixing in Zenker. By this technique it is possible to pick out the neoblasts
in ordinary sections, only very few other cell types being now and then mis-
taken for neoblasts, i.e. certain nerve cells. The methylene-pyronine method
may also give good results.

The typical "resting" neoblast, that is, not engaged in migrating to the
wound or in process of differentiation in the blastema, is roundish or more or
less spindle shaped, with an ovoid nucleus containing one to three nucleoli.
The cytoplasm is scanty, forming only about half the cell volume; it is
strongly basophilic due to abundant RNA occurring in a great number of
ribosomes, most of which are free with only a few bound to an endoplasmic
reticulum. The rich occurrence of ribosomes and of proteins containing
sulphhydryl-groups may be taken as a token of the neoblasts being a cell type
of embryonic character destined for morphogenesis. It is not known if these
cells have any other functions in the planarian body. It is perhaps significant
that a distinct nucleonema is present (Wetzel, 1961).

As to the origin of the neoblasts in the intact body, especially during regeneration, most authors think that they increase in number by mitosis,
although some, referring to the older literature, regard them as "dedifferentiated" cells (see Chapter 7). In a recent study Woodruff and Burnett (1965)
maintain that the neoblasts in the intact animal are continuously recruited
from intestinal basophilic gland cells. They showed photomicrographs of
intermediate stages between these two cell types. They further maintained
that during regeneration the numbers of neoblasts progressively increased as
the numbers of intestinal gland cells decreased and in the summary they
stated that "mitosis in neoblasts cannot account for the large numbers of these
cells in regenerating worms".

In view of the fundamental problems involved, the findings of these authors
definitely need strengthening by meticulous counting of the two cell types
during regeneration. I also think that the postulated transition of gland cells
to neoblasts should be reinvestigated by strict histochemical methods.

As to the embryonic origin of the neoblasts the informative papers of Le Moigne (1965a, b and 1966) should be consulted. His results strongly support the conception of neoblasts being the true regenerative cells.

7. *The epidermis*. This covers the external surface of the body as a one-layered epithelium. The cells are more or less columnar on the dorsal surface of the animal, cuboid or somewhat flattened on the ventral one. The nuclei are globular or ovoid with a somewhat irregular surface. Cilia are abundant on the ventral epithelium, rather scarce on the dorsal one. When undisturbed the planarians may be seen gliding smoothly and seemingly without muscular contractions on stones, water plants, on the bottom of petri dishes and so on, also sometimes at the water surface with their ventral surface attached to the water–air boundary. When disturbed the animals secrete abundant mucus by cells lying beneath the epidermis proper. This mucus is provided by the rhabdites deposited in epidermal gland cells; when the rhabdites are expelled they swell enormously in the water, often in a heavy mucoid mass which may prove a nuisance to the experimenter. The rhabdites are generally rod shaped, strongly acidophilic structures; they may crowd the dorsal epidermal cells so heavily as to blur the features of the cells themselves. They were investigated by Pedersen (1959) and Török (1958).

The epithelium is as already mentioned separated from the interior of the body by a basement membrane a few microns thick.

8. *Gland cells*. These are of several kinds and are found nearly everywhere in the planarian body. Lying just beneath the subepidermal basement membrane imbedded in the parenchyma are acidophil, basophil and rhabdite-forming cells (Pedersen, 1959a, 1963). Both acidophil and basophil glandular cells are found on the ventral surface of the animal; in Dendrocoelidae they may form glandular muscular adhesive organs ventral to the head. They are surrounded by acidophil cells.

The glandular cells often penetrate the epidermis extracellularly by a slender neck. Those interested in the cytology of the gland cells may consult Klug (1960).

9. *Sense organs*. The eyes, from two to six, are situated in the head: in *Polycelis* and related species they form a fringe along the sides of the anterior third of the body. The eye cells are of the inverse type, embedded in a cup of pigment cells. The black eye cup is usually situated medially in an unpigmented area, thus giving a cross-eyed impression, hence the old name *Planaria torva*.

Here the paper of Rölich and Török (1961) dealing with electron-microscopic investigations of the eyes of *Dendrocoelum lacteum* and *Dugesia lugubris* should be consulted.

Tactile sensory cells, chemo- and rheoreceptors are placed in several regions of the body, especially in "auricular grooves" on the sides of the head.

10. *Pigment cells.* The colour of planarians varies from species to species and often also between individuals. The colour is due to mesenchymal pigment, which may be contained in cells but usually occurs as pigment spots or vesicles. The newly formed regeneration blastema is devoid of pigment indicating that the neoblasts have not yet developed into fixed mesenchymal cells.

In most species the pigment giving the animal its external colour lies just beneath the epidermis.

CHAPTER 4

MORGAN AND CHILD

DURING the development of experimental zoology a decade or so either side of 1900, two names concerning planarian regeneration are outstanding: Thomas Hunt Morgan and Charles Manner Child. The work of these two men laid the foundation for the systematic scientific investigation of regeneration. Both were deeply concerned with finding the solution to the riddle of regeneration. They both failed, however, as we in our time are still failing, but their work brought forth a wealth of facts and stimulated other scientists to probe further. We therefore still owe a great debt to these outstanding investigators, whose names must be remembered with great veneration.

Morgan started his work on *Planaria maculata (Dugesia tigrina)* (1898). Only some of the more significant of his many results will be emphasized here because they are important in the formation of problems. He found that regeneration may proceed everywhere in the body except from the parts just in front of the eyes. This was not due to the small size of this part, because even smaller pieces, down to 1/279th, from other parts of the body may regenerate; Morgan concluded rightly that the non-regenerative power of the head tip was not due to lack of material. He observed that in all pieces enough material collects anteriorly to regenerate a head, the first organ to be formed, the next to form being the pharynx. If these two organs are too large in proportion to the future body, then the whole piece elongates, and thereafter a remoulding of the entire body takes place. How this is done is unknown. Morgan later (1900a) framed the word *morpholaxis* for this process, now called *morphallaxis*, etymologically a more correct term.

Morgan found that a narrow strip cut lengthwise often regenerates a head at the anterior end, occasionally two, hence the new longitudinal axis does not always run parallel with the original one. In these side-strips a new pharynx may arise in the old tissues. Morgan did not find any difference in the rate of head regeneration in pieces from the anterior and posterior part of the body. This is not surprising in this species which propagates also by fission, and Child has shown that although the ability to regenerate heads diminishes posteriorly, it increases again in the distal region where new heads of new individuals arise through fission. Here we encounter an example of the caution one has to exercise when using different species. One cannot generalize from one species to another concerning the rate of head regeneration along

22

the axes of the animal and we shall have to return to this question later, in Chapter 5.

Morgan's work (1900a) is indispensable for those who are interested in what marvels of regeneration planarians can perform. His general conclusions may be summed up as follows: he was in doubt as to how many of the changes in form during regeneration were due to the migration of material and how much to morphallaxis (remoulding) and to a great extent we are still ignorant in this matter. He did not like the expression "formative forces" because it does not tell us anything and is therefore unfruitful. He thought there must be something in the structure or composition of the old tissue which determined omnipotent cells to regenerate missing parts and generally these alone. His lengthy discussions on this topic are, of course, partly out of date, but none the less they provide good reading because they reveal sound and clear reasoning.

Planaria (Dugesia) lugubris is a hardy species used by Morgan (1902a) to probe further into regeneration problems. If a transverse incision just behind the eyes separates the anterior part of the head, this will often regenerate a head on the posterior surface, so forming a "Janus head", the new one with opposite polarity. Morgan suggested that this was not due to the influence of the brain because the incision lay anteriorly to it and the piece was devoid of brain tissue (however, this claim of Morgan's is not fully substantiated). Furthermore, he said that this phenomenon was not due to lack of intestine, because the piece still contained remnants thereof. In this matter Morgan touched upon the later much discussed reversal of polarity; we shall return to this problem later.

In his paper Morgan stated the very important fact that anterior pieces without ovaries and testes were still able to regenerate whole animals which, when mature, develop sexual organs; they therefore contain the keimbahn in some form, presumably in the neoblasts. In the same paper Morgan again emphasized his thesis: "The nature of the old piece is the determining factor in the subsequent growth in the new part." We shall see that this view is very different from that of Child's.

Morgan also states that if regenerating animals are not able to take up and digest food, the old parts are protected against extinction; they give material to the new tissues which therefore in all probability grow by cell division. This problem is an important one and well worth investigating more closely because it is connected with the problem of morphallaxis.

A detailed description of the ways in which segments produced by oblique cuts regenerate points towards later investigations concerning the time-graded regeneration field and heteromorphoses. Also the twinning produced by incisions along the median lines belong to this problem complex. Figure 17 shows one of Morgan's experiments: the animals were decapitated; a cut was made posteriorly in the median line, the one half was shortened by a transverse cut, the two halves were then held apart and both regenerated heads; if they coalesced only one head was regenerated and always the anterior one. This will also be discussed later.

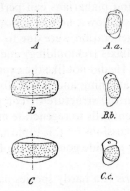

FIG. 17. Morgan's experiment. See text. (From Brøndsted, 1946, after Morgan, 1902a, *Arch. Entw. Mech.*)

Morgan was fascinated by the enigmatic "heteromorphic heads" (Janus heads). He later returned to the problem (1902b) using both *Dugesia lugubris* and *D. maculata (= tigrina)*. He stated that short pieces of *D. maculata* from the asexual productive region often produced Janus heads, whereas he had never seen such heteromorphoses in corresponding pieces from *D. lugubris*, a species in which asexual fission was unknown. Morgan soundly reasoned that this species difference was due perhaps to *D. maculata* being a species reproducing asexually, but he was not satisfied with this hypothesis and he further reasoned "—the fact that the heteromorphic head appears only in short pieces will probably also have to be taken into account...".

Morgan substantiated this view subsequently (1904a), again using *D. maculata*. He found that the shorter the segments the more frequently Janus heads appear. The development of a posterior head with reversed polarity was in his opinion not due to the influence of the anterior head; if this was cut away posterior ones did not develop. He combined his ideas about the development of Janus heads with those of his hypotheses about the determinative influence from the old tissues into a hypothesis of polarity suggesting that the larger pieces had a "stronger" polarity, hence a stronger inhibiting force in longer pieces concerning reversal of polarity.

It is worth while to cite Morgan's views in this matter (1904b):

> I have suggested tentatively that this means that in *Planaria maculata* the tendency is stronger for the new structure to become a head than a tail, and that when the influence of polarity is removed a head appears on each end of short cross-pieces. ... Can it be that there is a greater difference, chemical or physical, between the two ends of a longer piece, so that a stronger polarity is present?

In *Planaria simplicissima* Morgan found that the tendency to form tails in posterior pieces was stronger that the tendency to form heads, and "Janus tails" appeared when polarity was reduced or removed.

An example of Morgan's foresight appears in his investigations on *Dendrocoelum lacteum* (1904c). He found that a blastema was formed from the anterior wound surface in posterior pieces, but no regeneration of head followed (already observed by Lillie, 1900). Morgan stated that the lack of power to regenerate a head from wounds behind the pharynx cannot be due to lack of or lesser development of nerve cords, because these are adequate. He was unable then to find an explanation, as we are today.

Morgan (1905) again summed up his views on the causes of polarity by saying that it was an expression for gradation of material.

Together with Schiedt (1904) Morgan, experimenting with *Phagocata gracilis*, stated that in cross-pieces more supernumerary pharynges were formed in posterior pieces than in anterior ones. This was an expression of graded material along the main axis. This feature is still worth investigating more closely.

In his book on regeneration (1901) Morgan stated, besides other interesting facts, that fasting planarians regenerated more readily than non-fasting, but to a much lesser extent than non-starving, a feature which has been substantiated later by several authors and considered along biochemical lines. We shall return to this problem later (Chapter 15).

The great contribution of C. M. Child is to have put the regeneration problems on a physiological and semi-quantitative basis. His principal ideas are still under discussion. His hypotheses have called forth a wealth of investigations which have added substantially to our store of information. Child's book (1941), a monument to his, his pupils' and co-workers' investigations, is indispensable for every serious student of morphogenesis.

In a series of papers (1903a, b; 1904a, b; 1905a, b, c, d) Child investigated changes of form during regeneration, using *Stenostoma, Leptoplana* and *Cestoplana*. His general thesis was that normal locomotion by its stresses from contraction and relaxation determined the ensuing normal body form in the regenerate animal; this was so both whether the animal crept or swam. In one experiment Child hindered the animals from adhering by a gentle stream of water from a pipette. The controls which adhered to the substratum attained the normal elongated shape faster. He therefore concluded that locomotor activities influenced the shape. Here Child touched on one of the fundamental problems of morphogenesis: is motility necessary for giving the animal final shape during ontogenesis, and, if so, at what stage? We know from *in vitro* experiments in vertebrates that locomotion is necessary to give the finishing touch to the internal structure of bones. But in planarian regeneration I am not so sure that this idea holds: from my own observations I have noted that regenerating small pieces held in complete darkness do not move before the shape is complete. The problem certainly deserves a closer examination under strict experimental conditions.

Child developed his idea (1904c) that co-ordinate motor activity was for the greater part determined by the ganglia; he concluded that the significance of the ganglia for morphogenesis was mainly indirect through stimuli to the body parts, the function of which determined morphogenesis. He further concluded that regeneration from a given region was qualitatively complete in pieces both with and without ganglia, but quantitatively better in pieces with ganglia, and better still in pieces near the ganglia than in those taken from regions some distance from them.

It is a very interesting fact that Child found poor regeneration in pieces with one half of the ganglia removed and still poorer when the ganglia were removed completely. This finding pointed 50 years into the future to when Lender (1950) demonstrated the importance of the head ganglia for eye regeneration. The difference in interpretation is, of course, obvious.

Child criticized (1906c) Morgan's hypothesis for head-forming substances, saying that such a concept only led to a separation of form from physiology. That Morgan looked deeper into the problems the present times have demonstrated. I think that in these matters Child had a narrow viewpoint and a narrower concept of physiology. Morgan certainly included "head-forming substances" in his concept of physiology. In this connection Child's discussion with Driesch (1907) may be mentioned, even if it only has historical and philosophical interest.

Child (1910b) returned later to the problem of the influence of the nervous system on regeneration in Polyclads. He found poorer regeneration when head ganglia and nervous trunks were removed or nearly so. This agreed with Morgan (1905).

Turning to *Dugesia dorotocephala*, Child (1909) tentatively approached his later, and fuller, hypothesis of metabolic gradients. He found that anaesthetics modified or inhibited the regulative processes. However, the hypothesis continued to be that this was due to the control of influences from the nervous system and therefore secondarily upon locomotion. He (1910a) went farther into the problem by extensive experimentation with $1·5\%$ ethanol, $0·4–0·5\%$ ether and $0·025–0·0357\%$ chloretone. In these anaesthetics large portions of the body are necessary to get full regeneration, whereas small portions often remain without head; localization of the organs are often different and teratomorphs with only one eye or with supernumerary eyes frequently appear. Child concluded that all this might have been due to diminished uptake of oxygen, hence perhaps the same underlying cause basically of lowered metabolism.

Child (1906) observed that heads regenerated in various regions of the body were of different sizes, and he thought that this must be due to a gradation of physiological conditions along the main axis of the body; later

(1911a) he stated that the head was the first part of the body to be regenerated irrespective of how far removed from the original head of the animal the
piece was cut. If anaesthetics were used the metabolism was lowered, and in
these cases the regenerated heads were smaller, the distance from head to
pharynx less and the "dominance" of the head therefore weaker. He described the development of several forms of teratomorphic heads, and he again
ascribed this phenomenon to the lowered metabolism caused by the anaesthetics.

Without giving curves by way of illustration Child described the potentiality to regenerate heads along the main axis in such a way that the later concept of "head-frequency" was anticipated.

Subsequently Child returned to his favourite hypothesis that mechanical
stimulation increased the capacity for head regeneration, and he interpreted
this to mean that stimulation increased the rate or intensity of dynamic processes. He held that stimuli gave this potentiality to parts which normally do
not regenerate.

Some quotations from Child (1911b) should serve to give a clear picture of
his main theses, which play a major role later in literature on regeneration,
especially in that concerning invertebrates. He described the head region as
dominating regenerating bodies thus: "... it influences and determines the
processes and conditions in them. Each level of the body is to a certain extent
dominant over more posterior levels. The formation of the head in Planaria
... is the most fundamental, the most characteristic morphogenetic reaction
of the protoplasm of those species."

A cardinal point in Child's reasoning was that the power to regenerate or
not "may be the result of differences which are fundamentally purely quantitative rather than qualitative in nature". And: "... it is not too much to say
that the capacity for head formation is all that exists in the planarian egg, all
that is inherited". Although Child draws parallels to the animal pole of eggs
in his bold statements, it is curious that, notwithstanding his thesis, he stated
that "a new pharynx develops whether a new or old head is present or not".

Concerning the formation of new zooids Child (1911b) held that it "is the
result of physiological isolation of posterior regions of the body from the
dominant head region", and "we must regard the limit of dominance as
continually changing in the living organism". Here Child clearly expressed
slight doubt as to the complete validity of his thesis.

Later he voiced a very important concept in saying: "The group of cells
which give rise to a head in a piece of Planaria seems to me to constitute as
truly a germ as does the egg cell."

Besides a good deal of theorizing, Child (1911c) described numerous
differences in the size and shape of regenerating pieces in *D. dorotocephala*,
and he clearly stated that the "head-frequency" in this species was highest
anteriorly, falling off posteriorly until the hindmost quarter where it suddenly

rose again due to the existence of zooids in this region. On the other hand, the potentiality to regenerate a tail was greatest caudally, falling off anteriorly. In this paper Child says "—polarity consists essentially in a dynamic gradient or gradients along an axis", and in a way he is in agreement with Morgan in stating that a certain fraction of the gradient is necessary if complete animals are to arise.

Child's ability to write at length is pronounced in a paper from the same year (1911d). This paper is interesting in that he cautiously put forth his later clearly formulated thesis that oxygenation was the underlying factor in the gradation of morphogenesis.

In several papers (1912a, b; 1913b, c, d; 1916e) Child substantiated his concept of morphogenetic gradients as being due to metabolic rates. He used anaesthetics extensively as tools to strengthen this concept. After modern reports determining the physiological mechanisms involved in metabolism the papers are, of course, antiquated and somewhat naïve, but nevertheless they are of great historical interest. They are naïve in so far as Child in these earlier papers seemed to take for granted that susceptibility to anaesthetics served as a measure of metabolic rate, a notion which is encountered in several of his pupils' and co-workers' papers later (Chapter 17).

Child (1914c) expressed head-frequency by the formula x/y, where x represented the cells which build the head and y the rest of the fragment; therefore "a simple fundamental reaction system is the basis of development and inheritance in each species, race or individual".

Child's later papers appear in a period in which his ideas had called forth a broad discussion of Child's "Gradient Hypothesis", which will be discussed separately. The main thesis was that dominance and organization in morphogenesis is a gradation of a quantitative rather than a qualitative nature (1929b and 1941 and 1946).

CHAPTER 5

HEAD-FREQUENCY AND THE TIME-GRADED REGENERATION FIELD

THE term "head-frequency" is often encountered in the literature covering planarian regeneration. It is a well-established fact that a head regenerates with different vigour and at different rates from various levels along the longitudinal axis of the planarian body; usually the vigour and rate is highest just behind the eyes and tapers gradually posteriorly. A good example is given by Sivickis (1933): for the regeneration of eyes from transverse incisions at six consecutive levels in the body of *Dugesia lugubris* which were in hours: 104, 124, 139, 151, 167, 175 and in another experiment: 98, 117, 127, 133, 153, 184. Sivickis used only six specimens in each group. Dubois (1949c), using the same species, found the head-frequency curve given in Fig. 18 and H. V. Brøndsted's (1952) findings were the same.

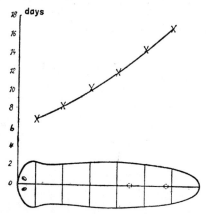

FIG. 18. Regeneration rate, in days, of head from various body levels of *Dugesia lugubris*. (From Dubois, 1949.)

The phenomenon of head-frequency was first systematically investigated by Child (1906a), and later (1910a, 1911a, b) he used this phenomenon to lay the foundation for his famous hypothesis of metabolic gradients (Chapter 17). The phenomenon has been extensively investigated in several species by pupils and co-workers of Child and others (Sivickis, 1923, 1930, 1931, 1933,

29

1934; Abeloos, 1930; Watanabe, 1935a, b, 1941b, c; Castle, 1940). Investigations of the head-frequency are numerous because the regeneration of a head is easy to follow under the dissecting microscope at a magnification of ×20–30, due to the easily recognized black eyespots. Although the study of head regeneration is of great importance I think that it has led ideas related to planarian regeneration somewhat astray, in the first instance by Child and his followers. There are many other regeneration phenomena in planarians

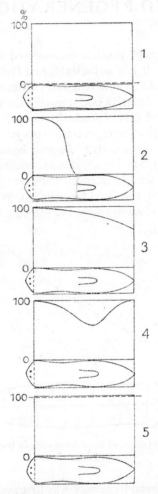

FIG. 19. Diagram of head-frequency in various groups of planarians. 1. *Belloura* group. 2. *Dendrocoelum* group. 3. *Phagocata* group. 4. *Dugesia dorotocephala* group. 5. *Velata* group. Dotted lines across the heads indicate the cuts separating heads from body before regeneration. Dotted lines in 1 indicate absence of and 5 absolute power of head regeneration. (After Sivickis, 1931, from Brøndsted, 1939.)

which involve neither head-frequency nor dominance of the head, nor even co-operation of the head. I am not sure that the hypothesis of metabolic gradients underlying the morphological gradient, starting with the head as the highest point, would have received the categoric form in which Child put it, if he and his collaborators had taken into account, for example, the curious regenerative power of *Bdellocephala punctata*. We shall revert to this matter several times.

Notwithstanding this qualification, the head-frequency curve is a phenomenon which must be known exactly in the species one uses as the experimental animal because errors in interpretation of results may then be avoided. Nearly every species has its own head-frequency curve as a species-specific character under specified conditions, among which may be enumerated temperature, degree of starvation and pH of the culture medium. In the conception of head-frequency I include here both the ability to regenerate a head from a given level and especially the rate of head regeneration from various levels.

The head-frequency curve is found in this way: the animals are incised at six, or preferably eight levels. It is essential to anaesthetize them in order to have them immobile, otherwise it is very difficult to cut at the intended level; even so it is often unavoidable cutting a little above or below the intended level. Experience has shown that twenty animals of each group cut at the intended level give a reliable average; the head, of course, is discarded. The twenty animals cut at the same level are put together in a Petri-dish, and the eight dishes kept in darkness at a constant temperature—in our laboratory we generally use 20 °C. Morning and evening the dishes are placed under the ×20 magnification dissecting microscope and the animals studied under constant illumination. When the eyespots are just discernible it is concluded that head regeneration will follow (Child, 1911b; Sivickis, 1930, 1931; Abeloos, 1930; Watanabe, 1935a; Brøndsted, 1942a). Following this one may plot the average numbers of heads and their rate of formation and so get the head-frequency curve. Sivickis (1930) worked out the head-frequency curve using five groups of the Triclads (Fig. 19). A tail-frequency curve may also be plotted, but with less exactitude; so far as I know it has only been done on *Bdellocephala punctata* (Fig. 20) (Brøndsted, 1939). From these studies it may be seen that the head-frequency curve falls off steeply just anterior to the pharynx, that is, no part of the body behind this level is able to regenerate a head, whereas at all levels tails are able to regenerate. It is of great theoretical interest that the two opposite curves need not necessarily be complementary; in *Bdellocephala* and *Dendrocoelum* the tail curve shows that regeneration is not confined to the forepart of the body as one might think if only the head-frequency curve was known. The phenomenon shows that we are concerned here not only with regenerative ability *per se*, but also with capacity for head formation; we shall discuss the problem later in different circumstances. Brøndsted (1942a) has shown in *Bdellocephala punctata* that the head-frequency curve is species-specific and already present in young animals emerging from their cocoons.

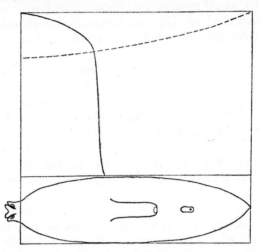

FIG. 20. *Bdellocephala punctata.* Diagram of head-frequency (full line) and tail-frequency (dotted line). (From Brøndsted, 1939.)

It is worth noting that the head-frequency shows certain anomalies in species propagating by asexual fission, especially when the initiation of a new zooid has begun; therefore caution is necessary when interpreting seemingly anomalous aberrations from the normal curve; examples were given by Watanabe using *Dugesia dorotocephala* (1935) and *D. gonocephala* (1941).

Brøndsted (1946) extended the notion of the phenomenon of head-frequency. The blastema formed at the anterior surface of a precise transverse cut is bilaterally symmetrical with a right and left eye appearing simultaneously.

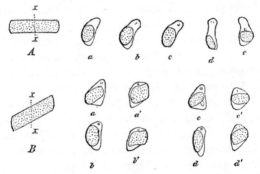

FIG. 21. See text. (From Morgan, 1902, after Brøndsted, 1946.)

We already know from other experiments that in some species almost every part of the body is able to regenerate a head. For instance, Morgan (1902) observed that if a transverse segment (Fig. 21) was severed in two by

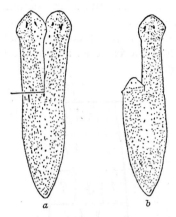

FIG. 22. Explanation in text. (From Li, 1928, after Brøndsted, 1946.)

a median cut each half regenerated its own head, the right eye appearing first in the right half and the left eye first in the left. Li (1928) produced a two-headed monster by a longitudinal median cut (Fig. 22) then he cut one two-eyed form away by a transverse cut and the blastema produced regenerated a new head.

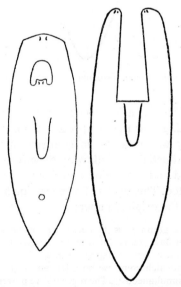

FIG. 23. *Bdellocephala punctata*. Explanation in text. (From Brøndsted, 1946.)

Brøndsted (1946), using *Bdellocephala,* decapitated the animal and cut the median part down to a level where head regeneration did not occur (Fig. 23); the animal thus appeared with two "arms", both of which made their own blastema in which eyes appeared after a considerable delay compared with the rate of eye-formation in the median part of an ordinary transverse incision at the same level; it is of importance to note that in the right "arm" the right eye appeared before the left one, and in the left "arm" the left eye appeared before the right.

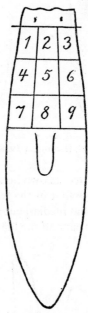

FIG. 24. *Dendrocoelum lacteum.* Explanation in text. (From Brøndsted, 1946.)

Dendrocoelum lacteum has a frequency curve nearly identical with that of *Bdellocephala.* Realizing that the head regeneration from other sites in the forepart of the body was retarded in comparison with the median part Brøndsted (1946) conducted the following experiment (Fig. 24): thirty animals were decapitated and the region anterior to the pharynx was cut into nine pieces as illustrated. The results of regeneration are given in Fig. 25, and may be quoted from the original paper:

> The ordinate of the graph gives the percentage of pieces having regenerated eyes in a given number of days (represented by the abscissa). The three thick unbroken lines give the regenerative power (expressed in eye-formation) of pieces 2, 5 and 8. These median pieces which comprise both left and right parts of the body always regenerate two eyes simultaneously. The regenerative power of these pieces coincided well with that of whole transverse sections, which have been employed to work out

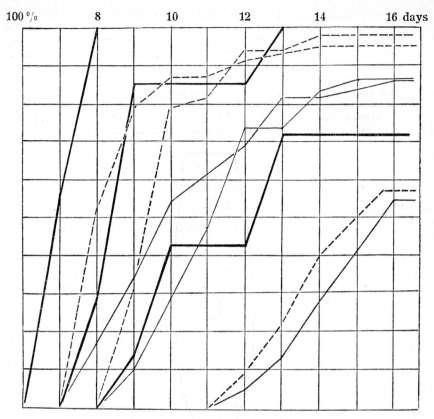

.FIG. 25. *Dendrocoelum lacteum*. Graph showing regeneration power of the nine pieces cut as seen in Fig. 24. Explanation in text. (From Brøndsted, 1946.)

the head-frequency curve of *Dendrocoelum* (Sivickis, 1931). This was of course to be expected, since the eyes on whole transverse sections originate close to the midline, that is, in the same tissue as that contained in pieces 2, 5 and 8. It will be seen that the 5-pieces start about 1 to 2 days later than the 2-pieces, and the 8-pieces again 1 to 2 days after the 5-pieces.

The 1-pieces behave of course in the same manner as the 3-pieces, because the body is bilaterally symmetrical also in respect to regenerative power (Brøndsted, 1952b). They were therefore calculated together, as were the 4 + 6 and 7 + 9 pieces.

1 + 3 (broken lines) are cut at the same level as the 2-pieces; nevertheless they start regeneration about one day later, and *then only with one eye each;* the second eye is regenerated after two more days (thin unbroken lines), and so the regeneration of symmetrical formations in the lateral 1 + 3 pieces are retarded 4–5 days.

The same occurs with the 4 + 6 pieces; they are retarded about 4 days in relation to the 5-pieces.

The effect is more pronounced with the 7 + 9-pieces; they are 8 days behind the 8-pieces from the same level of the body. ... I have tried to differentiate the experiments further by cutting the forepart of the body into 25 pieces, but without success; all the pieces disintegrated. This does not matter because there is not the slightest reason to suppose, that there should exist marked irregularities in the way of tapering of the regeneration field as revealed by the experiments in cutting the forepart into only 9 pieces. It must be borne in mind that this cutting in reality represents many variations in the sizes of the pieces, because it is impossible to make the cuts exactly equidistant from animal to animal. In some specimens the lateral pieces comprise somewhat more than one third of the body-segment, in others a little less; in these latter the regeneration is much retarded.

The results from these experiments prove clearly that there exists a time regulated regeneration field and a head-frequency field in Dendrocoelum... (Fig. 26).

FIG. 26. *Dendrocoelum lacteum.* Diagram of the head-producing time-depending regeneration field. (From Brøndsted, 1946.)

Starting from these experiments and from those of Brøndsted (1942), confirmed by Brøndsted and Brøndsted (1952) for *Dugesia lugubris* (Fig. 27), the conception of the existence of an inherent capacity for time-graded regeneration in planarians was established, and the significance for normal and abnormal regeneration emphasized (Brøndsted, 1954). The concept may be stated thus: the ability to regenerate a head varies in intensity and time not only along the main axis as expressed in a head-frequency curve, but also laterally in such a way that the ability to regenerate a head falls off evenly both posteriorly and laterally as illustrated in Figs. 25, 26 and 27. Regenerative activity is, of course, static in the intact body, but reveals itself when the body is wounded. The activity is very constant, the rate of regeneration being bound firmly to some as yet unknown structure in the body. The underlying factors will be discussed elsewhere, especially in Chapter 22. Brøndsted (1956) has conducted several experiments to strengthen this concept.

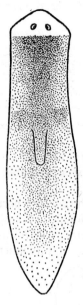

FIG. 27. *Dugesia lugubris*. Diagram of the head-producing time-depending regeneration field. (From A. and H. V. Brøndsted, 1952.)

Figure 28 gives one example: a median piece was exchanged with a lateral one; Fig. 29 shows that the median one in the lateral position still retained its ability to regenerate fully but that the lateral pieces transplanted to the

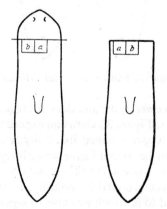

FIG. 28. *Dugesia lugubris*. Explanation in text. (From Brøndsted, 1956.)

midline, in spite of the new position, regenerated with its inherent slower rate and did not attain bilateral symmetry.

Figure 30 shows another example: the *a*-piece was transplanted to the hind-part of the body where the rate for head regeneration is much slower than that

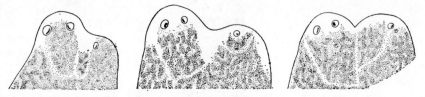

FIG. 29. *Dugesia lugubris*. Explanation in text. (From Brøndsted, 1956.)

of the forepart. At the level of the anterior surface of *a* in the normal position the eye-spots would have been seen after about 100 hours; this occurred with the transplanted piece *a*.

In Fig. 31, in the region of the anterior part of the piece *b*, eyes regenerate after about 140 hours; the rationale of transplanting *b* anteriorly to piece *a* was to see if the rate of regeneration of *b* was now accelerated; this did not occur, as piece *b* regenerated eyes at its inherent rate of 140 hours.

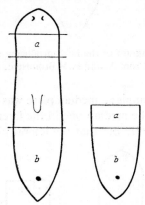

FIG. 30. *Dugesia lugubris*. Explanation in text. (From Brøndsted, 1956.)

We know that *Bdellocephala punctata* does not regenerate heads from parts posterior to the pharynx. Figure 32 shows an experiment performed to see if a head-regenerating forepart retained its ability when transplanted to the hindpart; this proved to be so and furthermore it regenerated a head at the same rate as if it were in its normal position, namely 90 hours. What is more, such a transplanted piece retained its polarity, as shown in Fig. 33. Here piece *a* was transplanted to *b*, which was able to regenerate a head, although at a rate slower than the more anterior piece *a*. In this special case the wounds

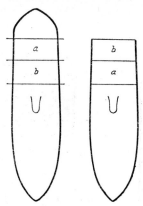

FIG. 31. *Dugesia lugubris*. Explanation in text. (From Brøndsted, 1956.)

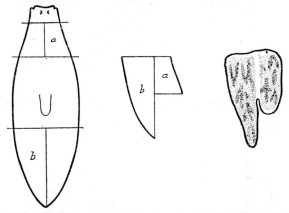

FIG. 32. *Bdellocephala punctata*. Explanation in text. (From Brøndsted, 1956.)

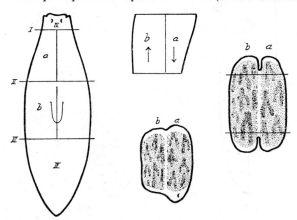

FIG. 33. *Bdellocephala punctata*. Explanation in text. (From Brøndsted, 1956.)

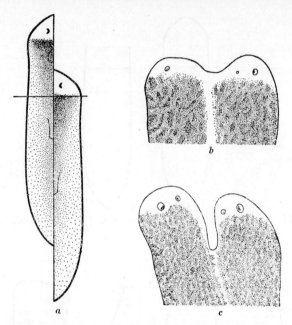

FIG. 34. *Dugesia lugubris*. Explanation in text. (From Brøndsted, 1956.)

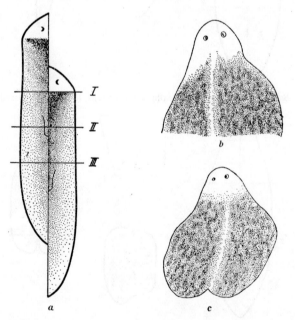

FIG. 35. *Dugesia lugubris*. Explanation in text. (From Brøndsted, 1956.)

closed so tightly that no blastemata were formed— a rather common feature in small pieces. Therefore two incisions were made as indicated in the figure; soon after the piece *a* regenerated a head with the original inherent polarity.

Figure 23 shows, as mentioned before, that in a right "arm" the right eye is regenerated first, and in the left "arm" the left. The same occurs if a decapit- ated animal is cut into two halves by a median longitudinal cut and separated; each half regenerates first the eye belonging to its side, and then a few days later the opposite eye appears. "Thus, in fact, the bilateral animal is made out of two one-sided animals acting together and controlling each other's ability to form a new symmetrical half. This example of inhibition may con- ceal a good deal of the morphogenetic riddle" (Brøndsted, 1956).

In a series of experiments two separate halves were transplanted together but transposed unequally along the main axis so the time-graded regenerative activities had to work together but at different levels. Figure 34 shows the result with a displacement of about one-quarter of the body length; the monster chimera was cut transversely as indicated and examined after 7 and 15 days. The transection has hit the right-hand part at a level with the higher rate of head regeneration and hence this was the first to produce eyes.

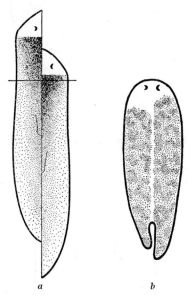

a   *b*

FIG. 36. *Dugesia lugubris*. Explanation in text. (From Brøndsted, 1956.)

Figure 35 shows the same degree of displacement; cuts II and III produced symmetrical animals, with the right eye always regenerated first. The two cuts crossed more posteriorly where the rate of head regeneration was slower; hence the left half had time to regenerate its own eye before it was inhibited from doing so by the right half, but both parts mutually inhibit one another from regenerating a symmetrical eye of its own (see also Chapter 10).

If the dislocation of transplantation was about half the body length, the transection crossed rather posteriorly in the left half where the rate of head regeneration was very slow, hence the right half completely dominated and suppressed head regeneration in the left.

If the displacement was only about one-sixth or one-tenth of the body length the result after decapitation was one symmetrical head with the right eye always appearing first as seen in Figs. 36 and 37.

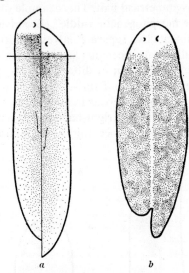

FIG. 37. *Dugesia lugubris*. Explanation in text. (From Brøndsted, 1956.)

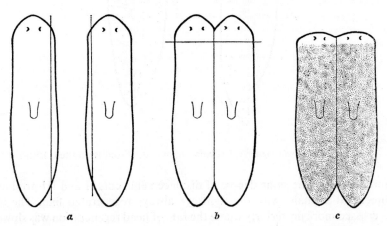

FIG. 38. *Dugesia torva*. Explanation in text. (From Brøndsted, 1956.)

In a series of parabioses between two specimens of *Dugesia torva* the following could be demonstrated: Fig. 38 shows two animals which had two narrow lateral strips removed allowing the animals to join. After decapitation both animals regenerated their own heads. If broader rims were removed, the specimens allowed to join and then decapitated, only one head regenerated, indicating that both left and right sections were able to integrate their activities (Fig. 39).

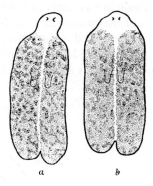

FIG. 39. *Dugesia torva.* Explanation in text. (From Brøndsted, 1956.)

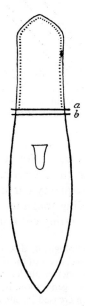

FIG. 40. *Polycelis nigra.* a, decapitation; b, cutting away the forthcoming blastema after varying time. (From Brøndsted, 1956.)

It is interesting to consider whether the time-graded activity was restored after regeneration, which is very likely, and, if so, even more interesting to know at what rate.

In *Polycelis nigra* the head-frequency curve is very like that of *Dugesia lugubris*. Figure 40 (Brøndsted, 1956) shows the basic experiment. By an incision (cut *a*) 115 specimens were decapitated just behind the eye-string. Fifteen individuals were kept as controls (group 0) for ordinary regeneration. Fifteen animals were cut again immediately at *b* 0·5 mm behind the first cut (control group I). The rest of the specimens were divided into five equal experimental groups I–V and allowed to regenerate for varying periods. Group I was subsequently cut at level *b* 2 days, group II 4 days, group III 6 days, group IV 7 days and group V 8 days after cut *a*. Figure 41 shows that the

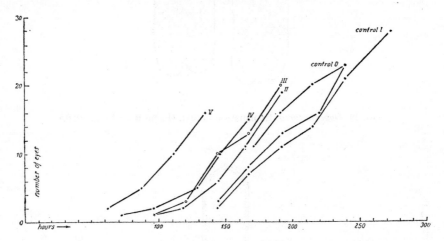

Fig. 41. *Polycelis nigra.* Eye-formation in various groups. Explanation in text. (From Brøndsted, 1956.)

longer the experimental groups were allowed to regenerate eyes before the second incision, the quicker was the rate of regeneration subsequently. This meant that the normal head-frequency curve grew more and more normal during regeneration; remembering that the second cut *b* was made in the severed tissue behind the decapitating cut *a*, it is

obvious that not only did a morphologically discernible regeneration occur at the level of cut *a* of the severed part, but in the rest of the tissues also there takes place a process of remoulding of structure and function underlying the phenomenon of the antero-posterior axis of the time-graded regeneration. In the course of 2–8 days, regeneration in the five experimental groups increased in rate at level *b* to that of a more anterior region and during the regeneration the morphological level *b* must have acquired the physiological properties characteristic of more anterior levels. This indicated that the morphological level *b* was physiologically adapted to the much shortened main axis which in turn had acquired the property of regular time gradation.

In another investigation, performed on *Bdellocephala punctata*, it was shown that six animals, which had been cut transversely, somewhat posteriorly in the time-graded area, did not develop eyes in the blastemata until 11 days later. A new transverse cut was made separating the blastemata from the newly formed eyes, and within 5 days eyes could be seen in the new blastemata in 5 of the 6 animals. This indicated that the time-graded activity was restored during reformation of the head by the blastemata (Brøndsted, 1956).

Chandebois (1955, 1957) found hemiheteromorphoses in the marine *Procerodes lobata*, varying in form and dislocation even in the returning of polarity, and she ascribes this to disturbances in the time-graded activity, because the rate of morphogenetic processes play such an important role. This extensive paper (1957) should be read in the original.

Teshirogi (1955a) found the same regeneration capacity in the Japanese species *Bdellocephala brunnea* (Ij. & Kab). He (1956b) transplanted the ganglionic region into postpharyngeal pieces of this species; eyes regenerated only in the graft, no production of eyes occurring in the blastema of the recipient.

Teshirogi and Ohba (1959) found the time-graded activity restored in *Bdellocephala brunnea* some time after blastema formation following an anterior incision. When the regenerating piece was cut away at the border of the old tissue a head was regenerated with greater speed than in the control; but if the cut is made 0·4 mm posteriorly to the border then no difference in the speed of regeneration between the experimental animals and the controls was observed. Eleven days later the regeneration rate was normal at all levels. The authors suggested that these findings were due to the migratory and mitotic activity of the neoblasts. In the region 0·3 mm posterior to the border in the old tissue several mitoses in neoblasts were found, and also several neoblasts in the process of migration. This activity was most pronounced 3–5 days after section. The authors were of the opinion that the blastema was formed by neoblasts derived only from the active part near the wound and not from distant parts of the old tissue, where mitotic activity was always infrequent.

It should be mentioned also that Pasquini, Guirardelli and Lesi-Massare (1955) suggested that neoblasts had a "funzione preminente" in the regeneration processes. Using small pieces ("dischetti") taken from the body of *Dugesia torva*, the authors did not find any "high-point" in the regeneration of these pieces. I think it difficult to evaluate this statement because no figures are given.

Teshirogi and Numakunai (1962), using *Bdellocephala brunnea*, transplanted tail pieces into prepharyngeal regions; here the graft developed tubular outgrowths with one or two pharynges and a mouth. These structures are, however, never induced in the host tissue.

The implications of the existence of a time-graded regeneration will be discussed in several places in this book, especially those pertaining to inhibitory forces (Chapter 10); possible factors determining the time-graded activity in the intact body will also be discussed in Chapter 22.

From this Brøndsted (1946) concluded:

> Let us imagine an accident: the animal is attacked and a part of the forebody with the head is snatched away by an enemy or other force. *The torn surface may be transverse, oblique, curved or straight; on the wound surface there is always a place where head-building takes place most energetically and in the shortest time.* This place, henceforward named the "high-point", is where the first reorganization of regeneration cells takes place.

CHAPTER 6

POLARITY

MORPHOLOGICAL polarity in living organisms has often been compared with crystalline structures in which a law-bound organization of atoms and molecules builds up individuals with clear boundaries in relation to the surrounding media, these often being without polarity themselves, e.g. air, water, and so forth.

It is relatively easy to determine the law-bound organization of atoms and molecules in, say, a quartz-crystal as modern investigations have revealed the order and strength of electrons binding the atoms together. In organisms, however, from the "simple" cell to highly organized forms such as flowering plants or human beings, we are met with such an overwhelming complexity of atomic, molecular and macromolecular combinations, that it is a formidable task to unravel how polarity is initiated and how it is held so tenaciously throughout the development of the individual from egg to the adult state. Thus the analogy between crystals and living organisms is very far-fetched.

In fact, the word polarity concerning living organisms covers a huge sum of complexities, although it is the very basis of organic morphology and we are only at the threshold of understanding this basis in biology. We can be sure that polarity in organisms is a result of an evolutionary process starting from the very beginning of the individualization of life, so it is unlikely we shall ever with certainty obtain the knowledge of how polarity began. We may, however, by indirect methods get some ideas. These methods utilize a combination of microscopic (including electron microscopy), biochemical and physiological investigations, and much experimental embryology is concerned with these basic inquiries. We like to believe that we have found how polarity starts in many organisms, e.g. the polarization of *Fucus*-eggs by light, the pH of media, by certain chemicals and so on; or perhaps we content ourselves by stating that some eggs get their polarity determined in the ovary, in some animals with the animal poles outwards, in others inwards, in relation to the lumen of the ovary. We observed that in *Dictyostelium* and allied forms certain concentrations of substance released from the cells determine polarity in the ensuing cell aggregate. In no instance, however, have we found the basic cellular clue in the cells themselves, namely, how cell mechanisms function so that factors outside the cell make it respond by organization of morphological polarization. It is reasonable to think that DNA-molecules,

47

themselves polarized, direct the polarization of the whole cell, and, starting from here, determine the polarization of the ensuing multicellular organism.

All these basic questions have to be solved before we really understand what polarity is in planarians also, but in these organisms we have a fair chance of getting valuable information at least about the coarser features of dynamics leading to structural morphological polarity. This opportunity occurs because it is relatively easy—in many cases at least—to divert polarity during regeneration processes produced by planned experiments.

With the word polarity in *Bilateria* we not only designate the antero–posterior axis of the animal but also imply dorso–ventrality and hence right–leftedness. It is necessary to stress this definition because it has not been given due consideration in some of the literature, e.g. when new heads in planarians arise as heteromorphoses, then it is pertinent to state not only that it is a head which is being produced but also how its entire polarity stands in relation to the rest of the body.

The literature of planarian regeneration is full of examples regarding reversal of the antero–posterior axis. The most spectacular are those of "Janus-heads" (Fig. 42) and "Janus-tails" (see also Chapter 4). They are deemed spectacular because a clear reversal of the main axis of the regenerate from the regenerant is more astonishing and more easily recognized than a slight lateral deviation of the regenerate from the axis.

Morgan and Child were deeply concerned with the problem of reversal of polarity (see Chapter 4). Using *Dugesia tigrina (maculata)* Child, like Morgan, found that short segments from the middle of the body may produce double heads or tails with opposite polarity. He ascribed this phenomenon as due to "physiological indifference". In *Dugesia dorotocephala* Child (1911b) described several types of Janus-heads and he concluded: "... polarity consists essentially in a dynamic gradient or gradients along an axis". He concluded, as Morgan did, that a certain fraction of the gradient is necessary for the production of a normal whole. But whereas Morgan ascribed polarity to some structural or chemical difference along the axis, Child definitely ascribed it to a gradient of metabolism. "The axial gradients in rate of reaction constitute the basis of polarity and symmetry in the organism."

In his survey (1929b) he rejected Morgan's views of polarity as a gradation of formative substances and again emphasized that the metabolic level determined polarity. In a very ingenious experiment Watanabe and Child (1935) found that even if short mid-segments often produce Janus-heads when cut out by two simultaneous cuts, such heteromorphs are not produced if the second cut separating the segment is made some time after the first cut; they held that in the latter case a stronger gradient of a metabolic type had already been induced by the first regenerate from the surface first exposed. This is precisely what we designated as the establishment of a "high-point" in Chap-

FIG. 42. *Dugesia gonocephala.* "Janus-heads" produced by demecolcine. (From Kanatani, 1958.)

ter 5, but it is evident that it tells us nothing about a pre-existing metabolic gradient in the intact animal.

So far it is clear that these experiments do not give criteria enabling us to discern between Morgan's and Child's hypotheses. We shall revert to the problem later when more evidence has been given.

We may here add that the two founders of the science of planarian regeneration seem to have confined their thoughts almost entirely to the polarity of the main axis of the planarian body neglecting entirely the dorso–ventral one. This may have been due to the fact that this latter axis was held more tenaciously in all experiments. I think that if Child had paid more attention to this axis he would have been more cautious in declaring polarity as the visible sign of a metabolic gradient.

Goetsch (1921, 1922, 1932) deduced from his experiments on the reversal of polarity that there were two sorts of regeneration cells: head-determining and tail-determining; however, this idea is only of historical interest.

Both Li (1928) and Kahl (1935) conducted several ingenious experiments on polarity, but their hypothesizing was too formalistic to be of value. Olmsted (1918) studied the regeneration of triangular pieces and concluded from his experiments that whatever regeneration may appear outwardly, the original polarity was not fundamentally transformed. Some grafting experiments by Santos (1929, 1931), Schewtschenko (1936a, b) and Sugino (1940, 1941) have been mentioned in another connection in this book. Here it should be remembered that the process of grafting produced an outgrowth from the host which may vary in organization from orthopolar structures through heteropolar ones to quite disorganized outgrowths. These latter are, of course, significant because they show that the inherent polarity may break down under certain circumstances.

Before proceeding to discuss more recent experiments concerning polarity in planarians I should like to remark that in species such as *Dugesia dorotocephala* and allied forms where propagation by fission is common, it is in my opinion very dangerous to draw general conclusions from experiments showing reversal of polarity, because such species have the power to form new individuals spontaneously throughout the body, hence they are more labile in their morphogenetic capacity. It would therefore seem that such species might reverse polarity with greater ease than species with a firmer morphogenetic composition. I therefore emphatically recommend the repetition of experiments aimed at altering polarity with species such as *Dugesia lugubris*, and especially with *Bdellocephala*, *Dendrocoelum lacteum* and other species which are without asexual fission. I have elsewhere (p. 31) emphasized the necessity to interpret the results of experiments in relation to the species used; it is inadvisable to generalize too freely in planarians! *Bdellocephala* and *Dendrocoelum* are, as said previously, genera which can only regenerate heads from parts anterior to the pharynx. Hence it follows that lack of head regeneration from hinder parts of the body cannot be taken as an indication that polarity cannot be reversed in such species as we simply do not know.

In recent works on planarian polarity the problem has been attacked bio-chemically and physiologically; among these papers some of Kanatani and Flickinger are outstanding. Kanatani (1958b), using whole specimens of *Dugesia gonocephala*, applied deacetylmethylcolchicine (demecolcine, Ciba) instead of colchicum, because it inhibited mitosis more effectively and was less poisonous. He observed the very interesting fact that even rather long segments cut out after the treatment regenerated bipolar heads ("Janus heads") (Fig. 43) and also that short segments from the anterior part of the animal developed bipolar heads more often than those from the posterior parts. Furthermore, on raising the temperature, the number of bipolar heads increased. If the treatment was carried out after cutting the animals into seg-ments, then bipolar heads occurred less frequently. Kanatani rightly pointed out that many authors (e.g. Child, 1911b and in other papers; Rustia, 1925; McWhinnie, 1955; Teshirogi, 1955 and others) have shown that bipolar structures are more frequent in postpharyngeal pieces (in species propagating with fission) when chemical treatment has not been applied. He suggested that the chemicals may possibly influence the nervous system. Kanatani (1958c) found a more retarded head regeneration in animals decapitated after treatment with demecolcine than if the chemical had been applied before decapitation. He was probably correct in suggesting that the higher susceptibility was due to the cutting, because it may be assumed that the chemicals penetrate more easily through the wound than through the un-damaged epidermis. Kanatani (1959a, b, 1960a) in later studies on the forma-tion of bipolar heads with the inhibitor demecolcine found that the respira-tion rate was raised in simple demecolcine-solutions, but unaltered if glucose was added. He also found that glucose abolished the power of demecolcine to produce bipolar worms and concluded that there must be some connection between these two features. He also observed (1959b) that the retardation of head regeneration by demecolcine was abolished by simultaneous adminis-tration of glucose or sucrose and (1960a) that segments from the forepart treated with demecolcine regenerated bipolar heads less freely when held anaerobically for 6 hours, or treated with KCN. In these circumstances tails are regenerated instead of heads. Kanatani suggested: "... that the rate of respiration actively associated with phosphorylation constitutes a limiting factor for the formation of bipolar heads", and: "... the reduced respiratory activity apparently leads to the production of a tail rather than a head at the posterior cut surface".

Beginning with the idea that differentiation was causally related to protein synthesis Flickinger (1959) used chloramphenicol (Chloromycetin) and deace-tylmethylcolchicine (Colcemid-Ciba), these substances being known to inhibit protein synthesis. In experiments with *Dugesia tigrina*, treated animals were

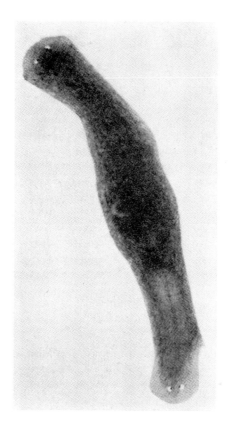

FIG. 43. *Dugesia gonocephala.* "Janus-heads" from a long anterior segment after decapitation, treated with demecolcine. (From Kanatani, 1958.)

Fig. 46. *Dugesia tigrina*. Explanation in text. (From Flickinger, 1959.)

incubated with $^{14}CO_2$ or ^{14}C and then cut into three, four, five or six segments of equal size; the carbon of the protein fractions was converted to CO_2, collected as $BaCO_3$ and estimated. Figures 44 and 45 show that there is an axial gradient of incorporation of ^{14}C. If the worms were cut into segments before treatment the same was found, that is, cutting did not modify the protein synthesis in the intact animal. Hence Flickinger demonstrated the possibility that the morphological antero–posterior axis is connected in some way with protein synthesis; it is worth noting that Flickinger found a slight increase in the rate of protein synthesis in cut-out tail pieces. Using antisera from rabbits Flickinger could not find any regional distribution of the main antigenic components, using anterior and posterior halves of the worms. He also confirmed Kanatani's findings that Colcemide caused bipolar head-formations, and that the presence of a head, in cut sections from Colcemide-treated worms, inhibited head formation at the posterior cut surface.

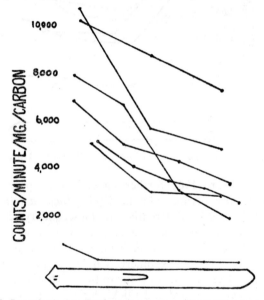

FIG. 44. *Dugesia tigrina*. Explanation in text. (From Flickinger, 1959.)

Flickinger further showed that chloramphenicol greatly retarded the rate of $^{14}CO_2$ incorporation and suggested that this phenomenon "reflects the synthesis of protein". He used a very elegant method which allowed the substance to influence the anterior end of the segments only: he imbedded the segments in thin slabs of 2 % agar so that the anterior end was free to dip into the solutions of chloramphenicol or Colcemide (Fig. 46). He concluded: "The evidence suggests that reversal of a gradient of protein-synthesis causes reversal of biological polarity in regenerating flatworms. Furthermore, it seems likely that this aspect of metabolic polarity, with its ancillary competition for nutriment, is a causal factor in the normal maintenance of polarity in regenerating planaria."

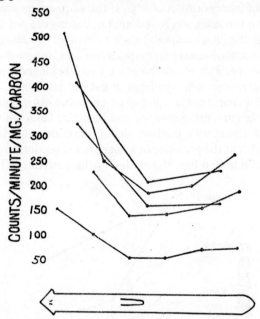

FIG. 45. *Dugesia tigrina*. Explanation in text. (From Flickinger, 1959.)

Flickinger and Coward (1962) further pursued these experiments, using the agar slab method, the experimental animal being *Dugesia dorotocephala*. The authors succeeded in getting differentiation of a secondary head at the posterior cut surface even though the original head was present anteriorly, but immersed in Colcemide. Even more remarkably, tail-pieces imbedded in agar, but with the anterior cut surface submerged in β-mercaptoethanol, sometimes transformed themselves into heads (four out of thirty-six cases).

It is only fair to cite Coward and Flickinger's later paper (1965):

> Earlier findings (Flickinger, 1959; Flickinger and Coward, 1962) have emphasized the relationship between protein synthesis and biological polarity, although it must be understood that the fact that one is not in a position to claim a causative role for protein synthesis may merely be a reflection of the gradient, or may be several steps away from the primary mechanism in the causal nexus of polarity determination and control.

So we see that these authors have grappled with the problem of morphological polarity by using physiological methods and they have succeeded in reverting the antero–posterior polarity at will. Both Kanatani and Flickinger were cautious in interpreting their results in firm terms and rightly so. Kanatani put some weight on metabolic rate measured by respiration. Flickinger was more inclined to think that competition for materials may be a factor in determining polarity. We are therefore lost in a maze of questions which have to be answered before we may see with confidence the links between morphological polarity, the underlying biochemical composition and physiological processes.

Many facts in planarian regeneration are still so bewildering that we must be very critical in our interpretation of experimental results. Let us take only one example: if morphological polarity were simply an expression of a gradient of metabolism (respiration) or a gradient of competition for materials— these two notions might, of course, be thought to have the same basis—then it would be hard to explain why, for example, *Dendrocoelum* or *Bdellocephala* cannot reverse the polarity of heads in the body parts behind the pharynx (Fig. 20) simply because they do not have the power to regenerate heads from these parts. I think that we have unfortunately concentrated too much on the two structures, head and tail, and not taken into account the susceptibility of various other parts and organs of the body for inducing factors. Therefore, neither gradients of metabolism nor competition for materials, e.g. protein, could be the only two factors in determining morphological polarity, even though they may probably play an important role during the regeneration processes; I think that induction and inhibition must first start the metabolic machinery. Therefore the basic answer to the riddle must be sought in some latent faculty residing in the intact worm. Several investigations (Chapter 17) have shown that no respiratory gradient was present in the intact body. Only on wounding are special regeneration forces released, and these seem to take the form of different metabolic processes, some of which Child and co-workers, Kanatani and Flickinger and others have pointed out as very likely to be parts of the whole sum of the physiological processes leading to regeneration.

In Chapter 23 we shall attempt to give a coherent picture of the whole problem-complex of planarian regeneration, by integrating the numerous pieces of evidence. Meanwhile, in this chapter on polarity, I, like Dugès, wish to point out that the same neoblasts may regenerate either head or tail, depending on the level of the transverse cut. Figure 47 shows schematically what I mean. If the cut is made at point *a*, the neoblasts in the neighbourhood regenerate a head, but if the cut is made at point *b*, the same neoblasts regenerate a hind part. Why? There must be some intrinsic factor in the rest of the body which determines the fate of the neoblasts.

I also wish to point out the interesting fact, detected by many authors but emphasized by Kido (1959), that after cutting, a border of disintegrated tissue 0·5 mm deep develops between the cut surface and the morphallactic processes leading, for example, to the formation of a new pharynx. I think close attention should be paid to this border of disorganized tissue as here we might be concerned with a nutrient medium, a sort of reserve, from which the neoblasts, and perhaps other cell types, derive substances necessary for the rebuilding processes. It might be that the subtle metabolic processes originating here may blur the metabolic picture gained by many experiments pertaining to polarity.

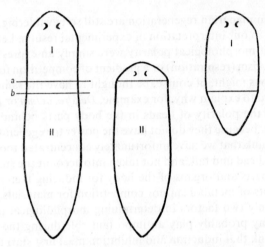

FIG. 47. Diagram of two cuttings close by one another. Explanation in text. (From Brøndsted, 1955.)

In an interesting paper Teshirogi (1963b) reported on transplantation experiments in *Bdellocephala brunnea* (see p. 45). When pieces were transplanted together the regenerates from the free surfaces were always in accordance with the polarity of the pieces. No regeneration took place if two sections were transplanted together with opposite polarity; but sometimes new tissues appeared from the line of union, e.g. a vertical pharynx or double-headed protuberances. He also described other interesting heteromorphoses.

CHAPTER 7

ORGANIZATION OF THE BLASTEMA

THE cellular events during regeneration may be considered under two head-ings, regeneration events in the blastema starting from a wound surface, commonly called *epimorphosis*, and regeneration events in the old tissues, commonly called *morphallaxis*.

When a planarian is wounded the body contracts around the wound. If the wound is not too extensive it is closed in the course of few minutes after the expulsion of mutilated cells. It is a matter of hours or days before the muscu-lar contractions relax revealing a whitish wound surface. If the wound is made by a clean cut, as is usual in experiments, the surface in the first hour or so is covered by a thin layer of flattened material hindering the escape of internal fluids and debris; later the formation of a layer of flattened cells is the first sign of the new epidermis. Just beneath this layer the blastema forms within a few days, the time differing from species to species and from region to region of the main axis. The blastema formed from a transverse cut even-tually in most cases takes the shape of a blunt cone, the first sign of epimorpho-sis of the lost parts. In the first few days the blastema is unpigmented even in coloured species.

Our problem is now: which tissues give rise to the blastema? As in most other regeneration phenomena we are concerned here with the problem of whether the blastema is made partly or exclusively of totipotent cells from an embryonic stock of body cells, the *neoblasts* (Randolph, 1897), or from al-ready differentiated cells in the neighbourhood of the wound.

It is quite clear that if the latter alternative is the right one we are faced with formidable problems; are specialized tissue cells able (1) to multiply so giving rise to the same cell types in the regenerated part of the body, and (2) to transform themselves into new cell types which have been removed by the mutilation, e.g. brain, eyes, other sense organs, etc.? If so, a de-differentia-tion of tissue cells must take place and then a new determination and differen-tiation ensue.

This old problem touches upon the very fundamental one: is the cytoplasm of differentiated cells capable of returning to an embryologic indeterminate state and thereafter—presumably under the influence of an intact genome and of surrounding old tissue and humoral influences—again be determined so that they differentiate into new forms of cytoplasm? I know this, that in regeneration we have no clear-cut evidence that such a thing may happen, and personally I do not think that it has been proven in ontogenesis either. Abeloos (1954) is of the same opinion. I think that Joseph Needham's comment (1942) that when a cell is physiologically differentiated "the doors are closed" is still valid. We know, of course, that practically every type of physiologically differentiated cell under certain circumstances, e.g. in tissue cultures, may dedifferentiate morphologically, that is, from a fixed state they may become amoeboid, losing some or all of their internal morphological characteristics. In the abnormal conditions after wounding, conditions very like those in tissue cultures may arise, giving cells opportunity to de-differentiate morphologically. It is a worthwhile task to probe into these problems with the help of such modern techniques as histochemical, electron–microscopic and cinematographic analysis. Whilst in vertebrates the problem has been handled well by Trampusch and Harrebomée (1965), in planarians this cellular problem is still under dispute, and we shall have to look at the available evidence shortly.

So it is not clear how the covering material is remade. Bartsch (1923) was positive, in his description of its formation, that the cells at the border of the wound coalesced to a syncytium which, as fine threads and pseudopodia in a state of solation glides over the wound surface, covering it in about half an hour. I personally have the impression that the covering material is first a kind of coagulated tissue fluid. This impression is much strengthened by the electron micrographs of Wetzel (1961) who interprets the very thin covering layer as flattened rhabdite material and his illustrations are convincing. In the following hours a true epithelium covers the wound. Bardeen (1902, 1903) held that the old epithelium from the neighbourhood of the wound effects the covering, possibly by cell division. Stevens (1901–2, 1907) and Stevens and Borin (1905) were inclined to the same opinion. Lang (1912, 1913) and Bartsch (1923) agreed but thought that regeneration cells later strengthened the new epidermis. Some of the older authors thought that amitoses occurred in the nuclei of the covering epidermis. In contrast to these authors Verhoef (1946) did not think that the epidermis took part in the formation of the initial covering of the wound and found no cell divisions, using the Feulgen technique. Dubois (1949c) also found no cell divisions, neither mitotic nor amitotic; she thought it probable that the new epidermis was made exclusively of neoblasts (Fig. 48).

Wetzel's (1961) electron micrographs suggested that the epithelial cells migrated from the base of the wound and after several days built up new epithelium, which was underlaid by a basal membrane only after 2–3 weeks. It is not clear from Wetzel's figures if the neoblasts also took part in the building of the new epidermis.

The problem of the initial covering of the wound is therefore still unsolved, and it can, I think, only be solved by using electron microscopy, using sections taken at short intervals during the first few hours. Perhaps phase-contrast microscopy in combination with time-lapse cinematography of small specimens might also be useful. This type of study should be done because the phenomenon needs to be elucidated as it touches upon fundamental cellular behaviour, although to hypothesize on similarities in wound closure in other animal phyla would be premature.

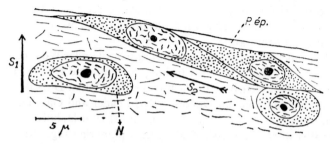

FIG. 48. *Dugesia lugubris*. Formation of epidermis by neoblasts, as conceived by Dubois (1949c).

In addition to the covering of the wound a large amount of mucus from the swelling rhabdites envelopes the animal.

As to the formation of the blastema itself both old and recent literature display contradictions. Most of the older authors thought that the blastema was made of totipotent cells of embryonic character, viz.: "Bildungszellen" (Wagner, 1890), "Stoffträger" with big nucleus and nucleolus (Lehnert, 1891), "Stammzellen" (Keller, 1894), neoblasts (Randolph, 1897). Morgan (1900c) was of the opinion that omnipotent cells took part in regeneration, as did Schultz (1902b). Although Stevens, and Stevens and Borin, in the above-mentioned papers, held that the new epithelium was made mainly of the old one, they thought that totipotent parenchymal cells migrating to the blastema were of paramount importance in the formation of new organs. Thatcher (1902) and Stoppenbrink (1905a) held that undifferentiated "Stammzellen" were the main regenerative cells, as did Child (1903a) for regeneration of the brain in *Stenostoma*. On the other hand, Child and Watanabe (1935) maintained that regenerated parts were formed by reorganization of the old parts and presumably not by undifferentiated cells. In contrast to this Beyer and Child (1930) and Wilson (1940) said that the blastema was primarily found in relation to the cut ends of the nerve cords. Lang (1912, 1913) and Steinmann (1908, 1925, 1932, 1933a–c) held, in varying degrees, that parenchyma cells were able to de-differentiate and, through intermediate stages, as transitory cells ("Übergangszellen") took part in regeneration. Steinmann conceded that all transformations were extremely complicated and difficult to interpret, and that all the complicated transformations occurred according to plan, presumably of an immaterial kind. He emphasized that this

3 PR

attitude was just as scientific as a mechanistic one. Vandel (1920, 1921a, b) and Prenant (1922) held views intermediate between those of Lang and Steinmann, whereas Curtis (1902, 1928, 1936), Curtis and Schulze (1924, 1934), Zweibaum (1915), Kenk (1924), Wilson (1926–7, 1941), Hein (1928), Abeloos (1930), Levetzow (1939) and Asperen (1946) more or less adhered to the neoblast hypothesis.

From this brief historical review it may be seen that the histological phenomena in planarian regeneration—as in planarian histology in general—are very difficult to unravel.

In the late nineteen-forties the Strasbourg school under Wolff made a great stride forward due to a new X-ray technique (Wolff and Dubois, 1947a, b, c; 1948a, b; Dubois, 1948a, b, c; 1949a, b, c; 1950; Stéphan-Dubois, 1951; Kolmayer, Simone and Stéphan-Dubois, 1960; Lender and Gabriel, 1961; Stéphan-Dubois, 1961). Among the cited papers that of Dubois (1949c) is outstanding; this work is a classic dealing with planarian regeneration, and is therefore indispensable for every serious student of this subject.

Her technique was as ingenious as it was simple (Fig. 49). By varying the position of a sheet of lead over the body of the planarian irradiation by X-rays could be effected either of the entire body or of certain segments according to the experimenter's plan. Dubois' irradiation of the whole animal gave

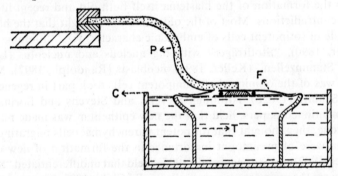

FIG. 49. Diagram of Dubois' irradiation experiments. P, lead screen the position of which may be varied; F, silk upon which the planarians are placed. T, frame carrying the silk, C, cup containing the nicotine solution. (From Dubois, 1949c.)

the same results as those already achieved by Bardeen and Baetjir (1904), Schaper (1904), Curtis and Hickman (1926), Weigand (1930) and Chevtchenko (1938), the irradiated animals dying during the course of several weeks (see also Chapter 19). But the studies of Dubois revealed that the neoblasts were more sensitive than the fixed parenchymal cells. It follows that a dose may be given which kills the neoblasts but not the mature differentiated cells; such irradiated animals do not regenerate if they are decapitated. Dubois next carried out a long and ingenious series of experiments, which should be read in the original. Here only the essentials will be mentioned. Figure 50 shows the effect of irradiation of the anterior part of the body, after which the animal was decapitated. If only cells from the immediate neighbourhood of the wound had the potential to regenerate a new head, then no

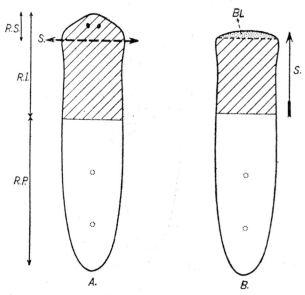

FIG. 50. Diagram of localized irradiation of a planarian. RS, head amputated along the line S; RI, irradiated portion; RP, lead-protected region, S, with arrow indicating the migration of neoblasts forming later the blastema, B. (From Dubois, 1949c.)

regeneration ensued. If any regeneration took place, a new head was made but only after a considerable delay. By varying the length of the part of the body irradiated head regeneration could be delayed proportionately to the length of the piece irradiated. Dubois' interpretation was that all neoblasts were killed in the irradiated part, but that neoblasts from posterior non-irradiated parts migrated forwards to the wound, formed the blastema and differentiated themselves to produce a new head. This interpretation was strengthened by the experiment given in Fig. 51. Here the regenerated head had the left eye appear first and later the right. The left part of the animal was irradiated over an area at a shorter distance from the wound of decapitation than the right, hence the neoblasts had a shorter distance to travel on the left side. Another experiment was also suggestive (Fig. 52). The right side of the animal was irradiated and three incisions were made in this part. The ensuing necrosis was first halted at the sites of the incisions due to neoblasts arriving at the wound, then in due course the neoblasts repaired the damage of the irradiation.

Dubois' theory was much strengthened by the work of Cecere, Grasso, Urbani and Vannini (1964). These authors, using *Dugesia lugubris*, labelled neoblasts with tritiated cytidine, and these cells were found again in the

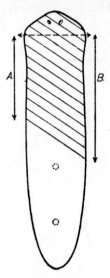

FIG. 51. Irradiation oblique. Explanation in text. (From Dubois, 1949c.)

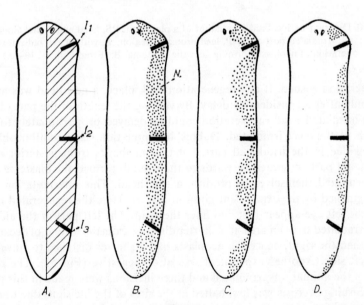

FIG. 52. Incisions made after half-side irradiation. Explanation in text. (From Dubois, 1949c.)

cephalic region after decapitation. If prepharyngeal pieces of labelled animals were transplanted to non-labelled ones the neoblasts migrated to the cephalic regenerating region (Fig. 53).

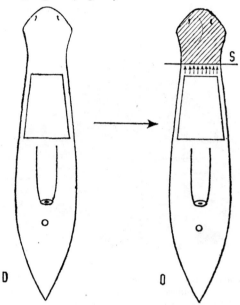

FIG. 53. *Dugesia lugubris*. D, labelled; O, non-labelled; S, decapitation level. Explanation in text. (From Cecere, Grasso, Urbani, Vannini, 1964.)

Lender and Gabriel (1965), using neoblasts marked with tritiated uridine, showed that neoblasts migrated to the wound and there build up a blastema in which new structures were differentiated.

Dubois dealt carefully with the cytological events. In agreement with several older authors, Dubois characterized neoblasts thus: «Les cellules de

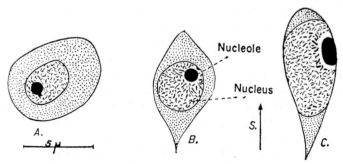

FIG. 54. Neoblasts. A, resting; B, C, migrating ones; S, arrow indicating migratory direction. (From Dubois, 1949c.)

régénération sont petites, arrondies ou faiblement ovoïdes, pourvues d'un gros noyau et remarquable par la taille de leur nucleole» (Fig. 54). And later: «Le blastème (normal animals) est formé uniquement de cellules de régénération qui se differencient à partir du 4e et 5e jour d'abord en muscles puis en cellules à rhabdites, cellules oculaires, etc.» The figures of Dubois (Fig. 55)

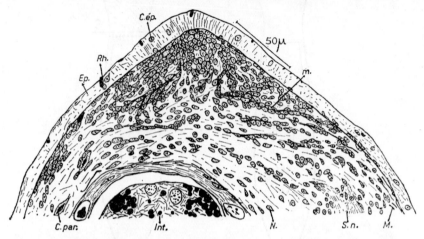

FIG. 55. *Dugesia lugubris.* Head regeneration after 5 days. The blastema filled with neoblasts. Ep, epidermis; C.ép, epidermic cells; N, neoblast; m, muscles formed by neoblasts; C.par, old cells of the parenchyma; S.n, nervous system; Rh, rhabdites; Int, intestine; M, muscles. (From Dubois, 1949c.)

showed the origin of the blastema filled with neoblasts. As to the differentiation of these cells the figures are not entirely convincing in themselves as they stand. A closer analysis of the differentiation processes with modern techniques is a future goal. The older literature is controversial as said previously.

Dubois seldom found mitoses in the blastema itself, but they were numerous in the neoblasts just behind the wound. Figure 56 shows the distribution of mitoses after decapitation of a normal animal; it is very interesting that the number of mitoses increases in the first 4 days but in such a way that a greater number was found somewhat posterior to the wound. It may well be that some stimulus from the wound travels posteriorly. Such a mechanism is, however, wholly unexplored, but the problem is so fundamental that it would be worth the considerable effort to unravel it. We are, of course, faced with the same problem during morphogenesis elsewhere in the animal kingdom; several hypotheses have been proposed, but it would lead us too much astray to discuss them here.

Figure 57 shows the number of mitoses in an irradiated animal; it will be seen that the appearance of the mitoses of the neoblasts in the irradiated anterior part were delayed, but that their number was greater. Dubois interpreted this as a migration of neoblasts from non-irradiated posterior parts and on arrival in the neighbourhood of the wound they undergo mitosis. The mitotic neoblasts found in the irradiated part did not therefore arise from there since no healthy neoblasts were found in the irradiated part during the

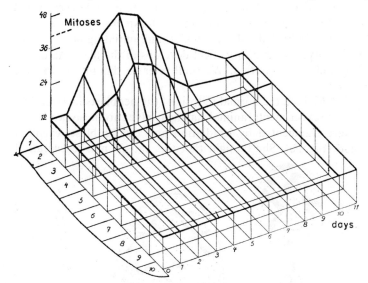

FIG. 56. Distribution of mitoses after decapitation of a non-irradiated *Dugesia lugubris* Explanation in text. (From Dubois, 1949c.)

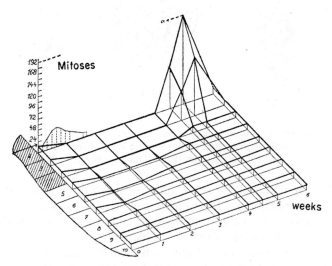

FIG. 57. *Dugesia lugubris*. Diagram of distribution of mitoses in an irradiated animal. Explanation in text. (From Dubois, 1949c.)

first 3–4 days. The increase in mitoses was, Dubois thinks, due to the fact that there were fewer neoblasts near the wound at the beginning of blastema formation; therefore mitoses were needed to augment the number of neoblasts in order to build a workable blastema. Dubois also stated that both the migration of neoblasts towards the new blastema and the mitoses declined when the differentiation processes in the blastema were in full activity. I think that this indicated a decrease of stimuli initiating both migratory and mitotic activity; an inhibitory effect develops in the blastema during determination and differentiation. We shall refer to this intriguing problem in Chapter 10.

Dubois described the normal mitosis of the neoblasts in *Dugesia lugubris*, her favourite experimental animal, having twelve chromosomes. The chromosomes were fragmented by irradiation and the neoblasts eventually disintegrated (Fig. 58). She thought that this caused non-regeneration of the

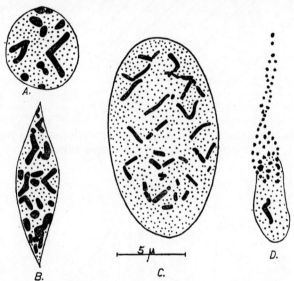

FIG. 58. Disintegration of chromosomes in neoblasts after irradiation. A, B, in metaphase; C, giant neoblast with fragmentated chromosomes; D, neoblasts in the course of complete disintegration. (From Dubois, 1949c.)

totally irradiated animals. It is not fully proven, I think, that resting neoblasts survive irradiation; if this were so, and no blastema were formed, then it seems necessary for neoblasts to go through at least one mitosis before they could form a blastema capable of differentiation.

Lender and Gabriel (1961) fully confirmed the findings of Dubois in *Dugesia lugubris*, and further stated that there were plenty of neoblasts even in worms starved for 2 months, and that they were still rich in RNA.

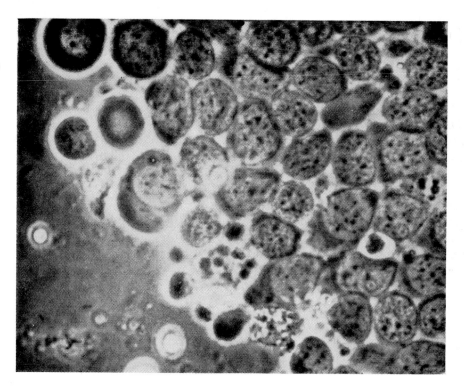

FIG. 59. *Phagocata vitta*. Living neoblasts from an early blastema. (From Pedersen, 1959.)

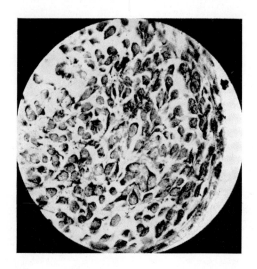

FIG. 60. Normal blastema 5 days after decapitation of *Phagocata vitta*. Carnoy fixed, stained with methyl-green pyronin (×940). (From Pedersen, 1958.)

Lender (1965) also stated definitely: "...whereas regeneration is completed by the neoblasts still in an embryonic state". Stéphan-Dubois (1965) again emphasized the notion of neoblasts as being the source of regeneration, and Le Moigne *et al.* (1965) also demonstrated in their electron micrographs that the blastemata of 48 and 60 hours old had a cellular structure. They found no sign of de-differentiation in the tissues in the neighbourhood of the blastemata.

This problem, together with others, was dealt with in Pedersen's important paper (1958). He had the fortunate idea of investigating the effect of tri-ethylene melamine (TEM) on morphogenetic processes in regenerating planarians. TEM is well known as a powerful anti-mitotic substance.

Most of Pedersen's experiments were made on the unpigmented *Phagocata vitta*. In the control animals he found that the wound (always a decapitation just behind the eyes) activated the migration of totipotent neoblasts towards the wound. He said:

> The migration is already evident on the day of decapitation and is very active for the next few days. Mitoses of neoblasts are evident the first day after decapitation. The rate of mitosis reaches a maximum on the third and fourth day. Mitoses may be found everywhere in the body, but as a rule they gradually become concentrated in the area adjacent to the distal part of the blastema after an initially more uniform distribution (Fig. 59).

And later:

> Serial sections show, however, that already on the day after decapitation a small blastema has formed. An increase in size is evident about the 4th–5th day (Fig. 60), and during the 5th–7th days growth of the blastema is most rapid. Further growth goes on after this, but at a reduced rate.

Pedersen's findings were in full agreement with those of Dubois. The variation in time of the cellular events was due to species differences in the rate of regeneration, this being somewhat slower in *D. vitta* than in *D. lugubris*.

In one experiment TEM was administered for 45 minutes at a concentration of 15 mg/l. Figure 61 shows the result, namely, that regeneration was strongly inhibited when treatment was given immediately after decapitation. Figure 62 shows the strong inhibition of mitoses by TEM. It is, however, very interesting that Pedersen did not find a significant inhibition of the migration of neoblasts. But inhibition of mitoses may nevertheless be deleterious to regeneration, e.g. a delay between the appearance of a hypoplastic brain and the induction of the eye-spots may be due to a diminished production of the eye-inducing substances (Lender, 1952, etc.), and the hypoplastic brain may be due to a scarcity of neoblasts because TEM had arrested mitoses. It therefore again seems—as in Dubois' experiments—that neoblastic cell divisions are a prerequisite to the formation of a healthy and vigorous blastema and it is tempting to draw a parallel with blastomere multiplication in early ontogenesis. When TEM was administered in the late phase of blastema formation no reaction was observed, indicating that TEM interfered neither with migration nor with the differentiation processes.

Pedersen ascribed another phenomenon to TEM: an interfering with the inhibitory forces in the time-graded regeneration field dealt with in Chapter 5.

Besides interfering with mitosis, TEM in stronger doses may break down the neoblasts in interphase, making them pycnotic or forming giant cells out

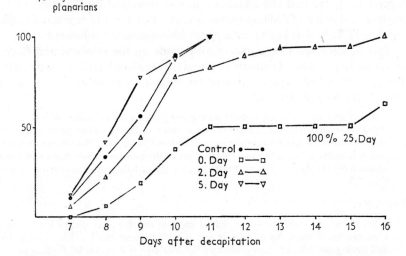

FIG. 61. *Phagocata vitta*. TEM treated. Explanation in text. (From Pedersen, 1958.)

Day after decapitation	Controls not treated			TEM-treated		
	Mitoses	*Nuclei*	*Per 1,000*	*Mitoses*	*Nuclei*	*Per 1,000*
Intact	60	3,012	0·79	··	··	··
,,	128	5,945	0·86	··	··	··
1st day.	238	6,011	1·58	44	3,379	0·52
,,	··	··	··	66	4,485	0·59
2nd day	228	5,847	1·56	47	6,072	0·31
,,	152	4,606	1·32	··	··	··
3rd day	241	3,978	2·42	22	6,005	0·15
,,	411	3,494	4·71	··	··	··
4th day	413·	3,721	4·44	8	4,807	0·07
,,	322	5,243	2·46	··	··	··
5th day	327	7,619	1·72	5	3,897	0·05
,,	178	4,407	1·62	··	··	··
6th day	148	4,531	1·31	··	··	··
7th day	138	5,045	1·09	··	··	··

FIG. 62. *Phagocata vitta*. Mitotic rate during regeneration. Experimental animals treated continuously with 5 mg TEM per litre. (From Pedersen, 1958, p. 317, table 1.)

of them, and then these cells may also break down. Furthermore, the basophilia of the neoblasts decreases. It must be noted also that TEM has no effect on the intensity of respiration. The influence of TEM on morphallaxis will be discussed in Chapter 8.

Whereas more recent investigations seem to show that the neoblasts form the blastema and differentiate themselves to form the missing organs and tissues, a few recent publications cast doubt on this conception.

Chandebois (1960, 1962 and 1965) says:

> Le parenchyme du blastème est édifié par la prolifération par mitose d'élements fixes du parenchyme, indifférenciés, apparemment agencés en syncytium. Les cellules basophiles accumulées contre la section dès les premières heures qui suivent l'opération ne participent pas à l'édification du blastème et sont considérées à tort comme des néoblastes. Elle se fragmentent par amitose et sont rapidémment détruites. Elles représentent peut-être des éléments de réserve des ARN.

And later:

> La découverte éventuelle de membranes cellulaires dans ce tissu ne modifierait en aucun manière les conclusions avancées dans ce mémoire.

In Chapter 3 we saw that Pedersen (1959a, b; 1961a, b, c) showed irrefutably with electron microscopy that no syncytium exists in the planarian body. It is impossible to demonstrate this with classical histological methods, so many misunderstandings exist in the older literature, including Lang (1912), Prenant (1922) and Bartsch (1923), which authors Chandebois regards as authorities in this matter. It is a curious fact that Chandebois in 1962 still talks of a syncytium, and it is perhaps this conception which is one of the fundamental sources of error in her interpretations. Chandebois used *Planaria subtentaculata.* Her principal technique was to make "empreintes", that is, pieces of tissue "légèrement dilacéré" were dried on the slide and stained with May-Grünwald, Giemsa. This technique is, of course, inadequate for histological investigations on planarian tissue because this very delicate tissue will break down or be distorted in other ways after a very short period outside the body, a feature which is very awkward for those experimenting with small pieces of planarians, e.g. in transplantation experiments. Chandebois' figures 1–5 show this inadequacy distinctly. Her figures 7, 8 and 9 are difficult to interpret; she described some of the cells as having "nucléoles pulsatiles", but from the text it is not clear whether pulsation had actually been observed, I am strongly inclined to think that the phenomenon was an artefact due to the inadequate technique. Chandebois' fig. 10 shows mitoses, and here the inadequacy of her technique is quite evident: some of the cells have their chromosomes irregularly dispersed due to the clumsy drying method.

Chandebois' fig. 6 represents a paraffin section through the blastema, and although she called it a syncytium, it is identical with the figures seen in Dubois' and Pedersen's papers, the blastema being made of neoblasts. Later, using the drying technique, Chandebois described curious cellular phenomena, amitoses of her cellules type 1 (= neoblasts); the figures show, as far as I can see, distinct pathological features again due to inadequate technique. Neoblasts are exceedingly frail and degenerate within seconds when not fixed properly instantaneously. Chandebois thereafter writes: «Des cellules qui se multiplient par amitose sont nécessairement dépourvues de potentialités histogénétiques et sont vouées à une destruction rapide.» And later: «Je pense donc que les cellules de type 1, après s'être fragmentées plusieur fois, subissent une cytolyse accompagnée d'une hyperplasie. J'interprète cette cytolyse comme un mécanisme de récupération des ARN par le syncytium.»

Chandebois' figures and interpretations are so unusual and involve so many completely new aspects of cellular activity that a deep scepticism is called for. Only new investigations made by suitable techniques can clarify the problems completely, and until then I prefer to regard Chandebois' findings, interpretations and unproven hypotheses as due principally to erroneous techniques.

The irradiation experiments of Dubois and others are, of course, no proof that the neoblasts in normal regeneration necessarily migrate long distances; on the contrary, several experiments have proven that the neoblasts situated even in short segments cut from the planarian body are sufficient in number to form blastemata both for head and tail regeneration.

Therefore the criticism of Flickinger (1964) concerning the idea of regarding neoblasts as the true regenerative cells is pointless, I think. Flickinger's experiments on which he based his criticism were as follows: *Dugesia dorotocephala* were incubated with $^{14}CO_2$; grafts from these worms were implanted in "windows" of non-labelled animals which were decapitated; Flickinger wrote: "The rationale of the experiment was that if movement of neoblasts to the blastema occurred, it would be detected by the presence of labelled cells situated in or near the blastema." This did not occur; on the contrary: "... the silver grains are largely confined to the graft and are not found in the area between the blastemata and the graft, nor in the blastemata". But, of course, we should not expect a migration of neoblasts from the transplant for two reasons: (1) neoblasts are available in sufficient number in the untreated worm for building up the blastema, and (2) the making of the window and the cutting out of the graft cause extensive wounding and therefore draw neoblasts from the host towards the transplant, and for the same reason the neoblasts in the graft stay where they are. Flickinger refers to Chandebois' paper discussed above. His reference to the paper of Brøndsted and Brøndsted (1961) was beside the point, since this paper dealt with an inability to regenerate heads in certain species from certain levels of the body, and hence dealt with the question of non-inductability.

Flickinger did not deal with cell types in the blastema, so his paper cast no doubt on the conception that neoblasts are the true regenerative cells in the blastema. This conception is supported in recent papers: Pasquini *et al.* (1955), Teshirogi and Ohba (1959), Wolff (1961) and Lender (1961), though a few, Bandi (1959) and Teshirogi (1962) hold that neoblasts are not the only source of regeneration in the epimorphic blastema. Wolff and Lender (1962) and Lender (1963) have given excellent reviews of the results of the French group, and Wetzel (1961) is very positive in designating the neoblasts as the source of the regenerative cells in the blastema.

Reese (1964), using the technique of autoradiography with tritiated uridine, was able to show that this compound was found in basophilic granules in the parenchyma and epidermis. The animals *(Dugesia dorotocephala)* were transected posterior to the pharynx and allowed to regenerate in the culture medium. Neoblasts showed little or no evidence of containing ^{3}H-uridine, whereas this substance was found in the epidermis and gland cells. The author concluded that these cell types were not regenerated from neoblasts.

It is of considerable interest to note that Tardent (1963) demonstrated that interstitial cells act as neoblasts in hydrozoa as the source of regenerative processes, and Stéphan-Dubois (1963) found neoblasts as true regenerative cells from the dissipiments of *Lumbriculus.*

Lindh (1957a) was reserved in his evaluation of the histological events in the regenerating head blastema in *Dugesia polychroa.* In describing the wound healing, he said that the cytological evidence was uncertain. All cells of the regenerating parts exhibited intensified mitotic activity (Chapter 18) and he wrote: "... but if not demodulating to 'neoblasts', most cells at the wound surface change their morphology and cytoplasmic relations and partake in the blastema formation, whereby they act as inducers...". He could not demonstrate significant neoblast movement in his investigations (Fig. 63).

It is pertinent here to cite the interesting paper of Stéphan-Dubois (1962) in which, with allusions to many animal groups, she broadly discussed the "germinal line" in relation to "totipotent" neoblasts, and concluded that in

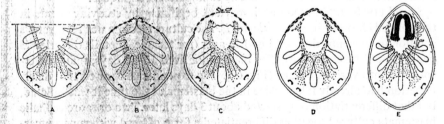

Fig. 63. *Dugesia polychroa.* Diagram of the morphogenetic events in a head-regenerating posterior part of the animal. A, transversal incision; B, immediate contraction of the circular muscles; C, central part closed (after 12–24 hours); D, 2 days of regeneration. Neoblasts replace the epithelial wound; E, 5–7 days: the blastema cells differentiate into tissue-bound cells. (From Lindh, 1957a.)

many cases the notion "germinal line" was without meaning—see Chapter 21.

I have just started an extensive study on the rate of organ formation at several levels of the planarian body. This study shows that the blastemata from all levels both anteriorly and posteriorly are made of neoblasts. A more detailed analysis will appear later.

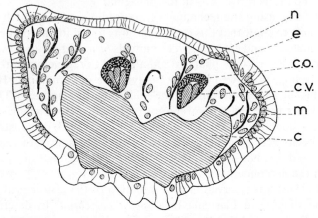

FIG. 64. *Dugesia lugubris*. Differentiation in a cephalic blastema cultured in modified Holtfreter solution. c, brain; c.o., eye cup; c.v., eye cones; e, epidermis; m, muscles; n, neoblasts. (From Catherine Sengel, 1960.)

I have dwelt upon the problem of regenerative cells in the blastema at length because I think it all-important in clarifying the problem. The elucidation of the biochemical processes during differentiation with the help of modern techniques can, of course, only be done effectively and reliably if the cellular morphology in the blastema is precisely known. I must again emphasize that most of the older literature on the morphological aspects of differentiation is rather unreliable due to imperfect techniques.

Sengel (1960, 1963) made an important beginning by studying events in the isolated blastema in *Dugesia lugubris*. The culture medium was Holtfreter's solution in concentrations of ½ or 1 in agar (for technical details the original paper should be consulted). The blastemata lived 10–28 days, many differentiating into various organs (Fig. 64). The experiments showed that determination had already taken place in the blastema at the time it was isolated, but that differentiation only started about 3 days later. Two or more cephalic blastemata cultured together differentiated a form of head with supernumerary eyes.

The experiment did not decide whether the blastema had already received instructions from the adult tissues before being cut off, or whether it was self-differentiating. I think, however, that the first is correct in view of the numerous indications given in the literature that the adult body is the determining factor for the overall outcome of the regenerate.

It is pertinent to relate here some experiments which on refinement might lead to a deeper understanding of the phenomena of a differentiation. The well-known phenomena of cellular behaviour after dissociation in Porifera (for review see Brøndsted, 1962) have encouraged some to investigate planarian tissues as far as possible with the same technique. Freisling and Reisinger (1958) cultivated restitution bodies in $\frac{1}{2}$ Holtfreter solution, but could not find any typical morphogenesis. Seilern-Aspang (1960) cultured small tissue fragments in a medium containing placental serum and found migration of cells without organization. Ansevin and Buchsbaum (1961) made a coarse brei of *Dugesia tigrina*; some were cultivated in 1/10 Holtfreter with planarian extract and they found that partial regeneration may occur in the restitution bodies. If cultured with chicken plasma and 2% agar restitution bodies were mostly inhibited, but sometimes migration of cells took place and sometimes morphallaxis and some regeneration was observed.

It is difficult to draw far-reaching conclusions from these experiments but they should certainly be continued, preferably with suitable culture media and with a technique whereby the single cell types might be isolated, e.g. after partial tryptic digestion. Those interested in this should consult the paper of Sengel (1964).

CHAPTER 8

EPIMORPHIC AND MORPHALLACTIC
HISTO- AND ORGANOGENESIS

THE literature dealing with the problems in this chapter is even more controversial than that of the origin of the cellular material forming the blastema.

Let us begin with a short historical review. Flexner (1898) held that the nervous system regenerates from the cells in the parenchyma. Bardeen (1902, 1903) thought that most tissues were regenerated from the same tissues in the old body and epithelium arose from epithelium by cell-division, the same being the case with nerves, muscles and intestine; but he admitted that large embryonic cells might also make a contribution. Schultz (1902b), on the contrary, held that the nervous system and genital organs were regenerated from embryonic cells. Stevens (1901–2, 1907, 1909) and Stevens and Borin (1905) were of the opinion that the epithelium was made mainly from the old epithelium, the intestine from both intestinal and embryonic cells. Child (1903a), for *Stenostomum*, held that the brain was regenerated from undifferentiated "Stammzellen", a view also held by Thacker (1902) and Stoppenbrink (1905a). In Chapter 7 we have touched upon the problem of the formation of new epithelium. Bandier (1936), after very careful work, stated that the intestine and the nervous system regenerated without the participation of regenerative cells, but that muscles were built both from juvenile muscle cells and regenerative cells together, and the pharynx exclusively from regenerative cells.

One would think that the cellular events in the *epimorphic blastema* were the most easy to investigate, but to this day no certainty has been attained even with the help of modern techniques. Wetzel (1961) studied stages of neoblast differentiation in the anterior blastema arising from decapitated *Dugesia tigrina*. His electron micrographs showed "an increase in the relative volume of the cytoplasm ... and the mitochondria are more numerous... The appearance of small elements of rough surfaced ER is noted. The elongation of these cells, and the presence of tiny tubules suggests that these pictures may represent young nerve cells." Here Wetzel was in agreement with Dubois (1949). Future investigators of the possible differentiation of neoblasts to true nerve cells should give close attention to the electron micrographs of Wetzel. In the blastema Wetzel describes "fairly extensive nerve bundles penetrating the blastema as early as 3 days after transection". Even after the work of Kolmayer and Stéphan-Dubois (1960), who claimed differentiation of nerves in the blastema on the tenth day after transection, it was not sufficiently clear whether these nerves were in reality differentiated neoblasts or outgrowths from the old nervous system. My personal view is that new nerve cells arise from neoblasts in the blastema and that they connect with outgrowing nerve fibres from the old nervous system. It is likely that nerve fibres from the old nervous system invade the blastema, as it has been suggested by Lender (1950, etc.) (Chapter 9) that the brain has an organizing capacity for eye-regeneration; there is also the fairly well-established fact that the nervous system has a neurosecretory activity (Chapter 13).

As to the gland cells Wetzel (1961) was of the opinion that such cells—infrequently found in the 3–7 days blastema—migrated with the neoblasts rather than differentiated *in situ*, but as he also thought that neoblasts might specialize into gland cells, this problem remains unsolved.

Wetzel's figures of the fine structure of muscle cells showed two sorts of filaments, one about 90 Å in diameter, the second from 130–290 Å. The arrangements seemed to be irregular. Wetzel found that "the fine structure of planarian muscle corresponds remarkably well to that of the translucent adductor muscle of the oyster". In addition to seeing the various morphological details described by Wetzel it should be possible by fixing tissue at short intervals to follow a differentiation process from neoblasts to muscle cells—that is, if such a process takes place, as described by some authors by light microscopy, e.g. Dubois (1949) and Seilern-Aspang (1960a) who observed cells *in vitro* using phase-contrast microscope.

In descriptions of *morphallaxis* in the older literature we often meet such notions as *de-differentiation, redifferentiation* and *histolysis*. Although Asperen (1946) thought that the different views "for a large part may be imputed to differences in behaviour between the different species", I am convinced that this is not so; the different ideas are simply due to the difficulties of elucidating the cellular events.

Morphallaxis is said to be induced by the histolysis which invades the whole body to a greater or lesser extent from the wound apparently depending on how much of the animal is to be regenerated.

In planarians the term histolysis is a debatable one and denotes only a general feature, i.e. an apparent lysis of tissues and organs, claimed by some authors as a more or less complete disorganization of most organs, by others as a mild lysis of the parenchyma. An exact description of the cellular processes, morphological, physiological and biochemical, is sorely needed. Even in a comparatively recent paper, Hauser (1956), describing histolysis in *Planaria alpina* after feeding, held that the intestine and some other organs were completely destroyed by a gradual digestion of the intestinal epithelium and other tissues, which filled the anterior of the animal as a sort of paste;

from this paste the intestine and other organs were reorganized, but the phases in this process are unknown. Perhaps there may be something in this story— we shall speak about involution later in Chapter 14—but it seems at present more likely that the findings of Hauser may be akin to the descriptions of Willier *et al.* (1925) concerning food uptake in which the interstitial cells (= parenchymal cells) engulfed the food content of the gastrodermal cells which were heavily vacuolized after feeding. However, the authors did not mention histolysis, although it must be admitted that the microscopical picture was so diffuse and contourless that it would be taken for cytolysis.

I have used this example because it shows how difficult it is to unravel the cytological events involved in food uptake; by simply using light microscopic techniques it is nearly impossible to solve the problem. Everybody who has looked at gastroderm in well-fed animals will concede that the cell borders are so intermingled with one another and with those of the parenchymatous cells that the cell membranes simply evade observation. The electro-microscopic investigations of Pedersen (1961) seem to have established that the cells preserve their individuality even if during intensive phago- and pinocytosis the cell membranes may disrupt repeatedly here and there.

In view of these observations it seems to me that we must be very cautious in our evaluation and interpretation of the various presentations made in the literature concerning cytolytic processes leading to morphallaxis.

The basic problem may be stated thus: given a transverse section, a small regenerated worm with all essential organs emerges. From this it follows that most of the fragments of organs and other cellular material in the segment must transform themselves into a complete although small worm with a whole set of new organs.

Two distinct principal possibilities present themselves:

1. All remnants of the various organ systems continue to be cytologically intact, and then by displacement and possibly by cell division build up the new form of the whole small animal. If this is so, it still remains to be elucidated how organ systems not present in the segment are regenerated, e.g. brain, sense-organs, pharynx, copulatory organs and so forth.
2. All old organ systems are dissolved by complete histolysis and the material utilized by totipotent neoblasts to regenerate a new worm.

Neither of these opposing principles seem clear-cut; on the contrary, both seem more or less intermingled, depending on the internal situation in the worm or segment; at least, most authors envisage intermingling occurring in morphallaxis, although a few hold that the first principle is the only one at work or at least predominating. Let us give some examples:

Steinmann (1925, 1926) stated that de-differentiation of certain tissue cells took place in such a way that they transformed themselves into bipolar "Wanderzellen". Whether such cells were really neoblasts cannot be seen clearly from Steinmann's work; I personally think so, but still unanswered is the question: what becomes of the "de-differentiated" specific tissue cells? It is, in my view, highly improbable that Steinmann is right in claiming neoblasts to have arisen from de-differentiated tissue cells.

In opposition to Steinmann, Stevens (1901–2) and Thacker (1902) held that embryonic parenchyma cells regenerated new organs. Clearly neoblasts are involved in this. In addition Stevens was of the opinion that morphallaxis was also due to contraction and elongation of the old tissues without true migration of cells. This must then involve a displacement of already existing tissue cells, e.g. muscle and nerve cells. Bardeen (1963) held that highly differentiated tissues were destroyed unless they could be of direct use in rebuilding regenerating organs. In Rhabdocoeles, Hein (1928) embraced principle 2 in saying that regeneration was almost exclusively performed by neoblasts, only the epidermis taking part in rebuilding the new epidermis. Bandier (1936) approached this interpretation by saying that only the intestine and nerve cords regenerated in morphallaxis without involving neoblasts; otherwise according to Bandier the neoblasts were the basis of morphallaxis, and the brain was probably also regenerated by neoblasts in contrast to the nerve cords. Abeloos (1930) observed that morphallaxis was a process starting from the epimorphic regenerative blastema and proceeding from there posteriorly; from this it may be deduced that morphallaxis was due to neoblasts alone; but Abeloos seems not to have made detailed histological investigations. In Annelids, morphallaxis does not seem to be "induced" by an epimorphically established bud (Abeloos, 1965). Watanabe (1948) surveyed the different morphallaxes taking place in different segments of the body of *Polycelis sapporo*, but again the histological picture was not clear.

As stated previously morphallaxis is related to the phenomenon of involution. Castle (1928) described in *Planaria velata* the encystment of asexually produced animals. He found that the cysts were transformed into mucoid capsules in which complete de-differentiation of all organs took place giving a completely disorganized material with "... no morphological evidence of the original axes of the worm other than the difference between the dorsal and ventral surfaces...", and: "The original axes probably persist through the processes of fragmentation and encystment." When the animal again emerged as a normal worm the first organs to appear were gut and cerebral ganglia. Castle stated that a complete histolysis of all internal organ systems preceded fragmentation of the worm, leaving only a syncytium with few nuclei. Castle found that well-fed animals and a rise in temperature brought about encystment. If this picture of involution should be confirmed by investigations using newer techniques, then involution processes are of type 1, and true histolysis might well also occur during morphallaxis. But again we are without exact information about the salient question whether syncytial phenomena really exist, or whether, as I personally think, cell boundaries are still present.

It is quite another thing that phagocytosis undoubtedly occurs using up certain cells. This must be the case during starvation, for instance, which will be dealt with in Chapter 15. Most probably such phagocytic processes are also at work during morphallaxis.

The literature appearing after my review (Brøndsted, 1955) does not fully elucidate the cellular events during morphallaxis, even though positive claims have been produced. Among the more recent papers dealing with morphallaxis are the outstanding papers of Kido (1959, 1961, I and II).

Kido, using *Dugesia gonocephala,* made a special study of the regeneration of the pharynx in transected pieces. He cut out a transverse prepharyngeal, pharyngeal and postpharyngeal section, the latter with and without tail-tip. His aim was to see how the formation of a new pharynx occurred at the three levels. Figure 65 shows schematically his interpretation of the events in the prepharyngeal segment. The wound contracted as usual, and the two lateral nerve cords bent somewhat towards one another; from the posterior ends of them outgrowths grew posteriorly. It was not clear whether neoblasts accompanying the nerve cords regenerated these, or if they grew posteriorly without the participation of neoblasts. The intestinal tract retracted from the cut surface, and Kido claimed that "a syncytial tissue" appeared behind the intestine. The "syncytial tissue" divided itself into two lateral branches and

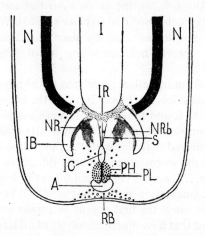

FIG. 65. *Dugesia gonocephala.* Diagram of pharynx formation during regeneration in a prepharyngeal piece. A, atrial primordium; I, original intestine; IC, primary cavity; IR, bridge of syncytial tissue derived from original intestinal tract; N, original nerve cord; NR, mass of regenerating nervous tissue; NRb, regenerating nerve branch; PH, pharyngeal primordium made by neoblasts; PL, caniculus from the primary cavity to the atrial cavity; RB, accumulation of neoblasts in the posterior blastema; S, "streaming pathway". (From Kido, 1961b.)

"comes in contact with the anterior cut ends of the old intestinal branches which happen to be there". In the middle of the "syncytial tissue" a "streaming pathway" grew posteriorly towards the cut surface, where disintegrated tissues blocked up the wound, and where neoblasts gathered to form the

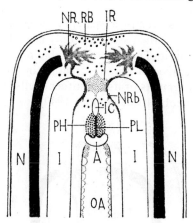

FIG. 66. *Dugesia gonocephala.* Diagram of pharynx formation of a pharyngeal piece after removal of original pharynx. Same abbreviations as in Fig. 65. (From Kido, 1961b.)

regenerative blastema. Later the distal end of the "streaming pathway" formed a primary cavity, which was then surrounded by neoblasts. These neoblasts soon developed into the pharyngeal primordium. Kido held that the newly formed lateral intestinal tracts were a remoulding of the old intestinal branches—up to that point without help from neoblasts.

In the pharyngeal piece (Fig. 66) the old pharynx was extruded and lost, a large part of the atrium only being preserved. The surface of the wound contracted strongly and the atrium retracted considerably, the anterior end disintegrating, and the remaining cells in the anterior part of the atrial wall tending to form a syncytium, the nuclei of which were unstainable with haematoxylin, only the nucleoli being eosinophilic; this tissue thus bore a striking resemblance to the syncytial tissue of intestinal origin. Necrosis was seen in some of the cells in the remaining part of the atrial wall. In the syncytial tissue of the anterior atrial part a cavity was produced, communicating through a canaliculus with the original atrial cavity. Neoblasts were found scattered near the cut surface and a syncytial bridge appeared from the cavity from the cut ends of the lateral intestinal tracts. From the middle of this bridge two projections emerged in both anterior and posterior directions. The posterior projection united later with the wall of the cavities in the syncytial tissue of atrial origin. The pharyngeal primordium was made of neoblasts accumulating around the canalicular part of the pharyngeal primordium, which according to Kido arose from the intestinal tissues. Later a new nerve appeared running transversely in front of the atrial and pharyngeal primordium, and the canaliculus passing through it developed into the pharyngeal lumen.

In the piece posterior to the pharynx (Fig. 67) the nerve cords also bent towards each other due to contraction of the wound. The old epidermis spread over the wound as a semi-transparent thin, syncytial tissue. Considerable histolysis occurred in the tissues not only near the wound, but also in the regions distant from it. The first remarkable phenomenon in the histolysis was the disappearance of fibrous structure in the mesenchymal tissue and the migratory movements of its constituent cells. These cells seemed to be identical with the neoblasts of Dubois (1949).

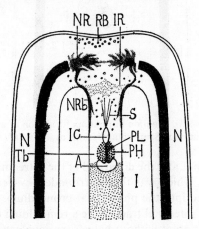

FIG. 67. *Dugesia gonocephala*. Diagram of pharynx formation in a postpharyngeal piece. Same abbreviations as in Fig. 65. (From Kido, 1961b.)

It is of considerable interest that intestinal tissue was injured by cutting

not only at the cut surface but also in parts distant from it. The tissue of the inner intestinal layer disintegrated and sometimes the constituent cells were scattered and the cells of the outer layer adjacent to the cut surface dispersed. The cytoplasm of these dispersed cells is stainable with haematoxylin and eosin, but their larger nuclei are dye resistant, except for the frequently found eosinophilic nucleoli. These cells make the syncytial tissue.

According to Kido, neoblasts build a "syncytial" tissue connecting the two original nerve cords. "It is sure that the syncytial tissue is made of regenerating cells derived from the original nerve cords." It is not clear whether Kido thought that the neoblasts were derived from the nerve cord cells themselves or from the neoblasts surrounding the nerve cords, which has been suggested by several authors and by Kido himself (1957) as in the intact animal the nerve cords are surrounded by numerous neoblasts.

Kido thought it very likely that the accumulation of neoblasts near the cut surface was the first step in the formation of the regenerative blastema. Fine dendriform processes from the neoblasts extend towards the cut-surface of the piece, and others towards the end of the main intestinal tract. But Kido held that: "from the medial corner of each healed cut end of the intestinal tract a band of deeply eosinophilic, and slightly basophilic, tissue appears. This tissue seems to be derived from the intestines." These cells were slightly larger than the neoblasts and contained fewer granules; it cannot be discerned from Kido's observations if these cells were in reality neoblasts under differentiation as no mitoses were found among the neoblasts forming the blastema, though a few were found near the cut ends of the nerve cords.

The original lumen of the intestine then began to invade the intestinal bridge of the syncytial tissue, and from the middle of the bridge an anterior and a posterior projection grew out; the latter Kido called a "streaming pathway" which invaded the future pharyngeal sac.

Just as in the pharyngeal piece a slender nerve branch occurred within the blastema and connected with the cut ends of the two original nerve cords; Kido said that: "Presumably the cephalic ganglion will arise from this new cord." He thought that neoblasts sprouted out a slender nerve branch along the inner side of each lateral intestinal branch.

Later a "primary cavity" appeared within a knob of the syncytial tissue situated at the end of the "streaming pathway". At the posterior part of the "primary cavity" a heavy mass of neoblasts accumulated forming a pharyngeal primordium. There were many histological sections which demonstrated the continuous distribution of these cells from the blastema to the cavity along the "streaming pathway" as well as along the inner sides of the later intestinal tract. This is very interesting because it indicated a clear connection between the nerve cords and neoblasts along the branches of nerve fibres growing down along the same paths. Later the primary cavity evaginated into the pharyngeal primordium as a slit, the distal end bulging into the secondary cavity behind the pharyngeal primordium. Thus, the cavity formed the atrium later. This description corresponds closely to the phenomena described by Bandier (1936). Kido continued: "... the anterior projection from the syncytial tissue of intestine enters the blastema and will constitute a median intestinal tract in the prepharyngeal region".

All this happened 24 hours after cutting and a normal organ formation developed after only 48 hours. A fibrous structure within the mesenchyme soon became distinct. Three to four days after cutting, slender muscle fibres were found in the blastema and in the new pharyngeal wall. Simultaneously, the posterior tip of the atrial wall began to extend towards a ventral body wall, finally opening on the outside as the mouth opening. Kido emphasized that the pharyngeal formation took place only in relation to the anterior cut-surface.

It is well worth noting that the new pharynx always arose 0·5 mm (Kido, 1959) from the anterior cut-surface, and certainly corresponded to the extent of the disintegration of the mesenchymal tissue between the lateral intestinal tracts. This was verified by Okada and Sugino (1937) and Okada and Kido (1943). If the postpharyngeal piece in *Dugesia gonocephala* was less than 0·5 mm, the piece lost its polarity and was frequently capable of developing a head in the reverse direction. This very interesting feature was touched upon in Chapter 6 on polarity. Kido emphasized that the primary cavity was derived in every case from the intestinal tissue. He assumed that the cells of the primary cavity of intestinal origin might induce the MR-cells (= neoblasts) to develop the pharynx. This is in accordance with Sugino (1938), and it will be discussed in Chapter 9.

Many investigators have demonstrated that cutting releases factors which induce neoblasts to migrate to the injured place; but Kido emphasized rightly that neoblasts also migrated to places distant from the wound and in his experiments to a place where they formed the pharynx; in this case therefore other inductive forces must have existed (see Chapter 13, where the role of the nervous system will be discussed).

In contrast to many other authors, especially Bartsch (1923a, b) and Dubois (1949), Kido held that neoblasts were only able to regenerate mesodermal tissues. Further, Kido found only very few mitoses, in contrast to several authors' findings. He said the cells participating in the regeneration must be the cells themselves coming from the old tissues. He further emphasized that in his material the pharyngeal formation by neoblasts took only 24 hours; this is perhaps the reason why he found so few mitoses, because mitoses in material forming other organs, e.g. heads, become numerous later, reaching a peak on the 3rd and 4th day, so we again face the problem of the "time-scale" of organ formation during regeneration.

Another very important problem raised in his work is whether Kido's "syncytial tissue" is a reality; it seems to me dubious, and the problem may only be solved by the use of electron microscopy. Perhaps some of the "syncytial tissue" is simply an indication of histolysis. I think that wound formation brings about disintegration of cells in the neighbourhood of the wound and from this autolysed tissue a mass of material is made available from which regenerative cells derive nutriment for further activity. We shall refer to this problem later in Chapter 23.

Kido (1961b) divided pieces of *Dugesia gonocephala* longitudinally in the mid line. Pharynx formation was almost identical with that after transverse section, but much delayed. He also inserted cephalic pieces dorsally into the postpharyngeal region just under the dorsal epithelium of the host. From the cephalic graft new nerves sprouted uniting with the nerves of the host. Along these nerves neoblasts migrated towards the portion between he two lateral intestinal tracts which had been destroyed by the transplantation. A primary cavity was formed within the bridge built between the destroyed ends of the two lateral intestinal tracts of the host. Behind this the pharyngeal primordium was formed by accumulating neoblasts, a second cavity behind developing into an atrium.

I have reviewed and discussed Kido's papers at some length because in my opinion they give the best information so far of morphallastic processes concerning cell and organ formation.

Sengel (1956) made the important discovery that the pharynx may be regenerated both by morphallaxis and by epimorphosis. In a prepharyngeal piece the pharynx was regenerated by morphallaxis, in a postpharyngeal piece by epimorphosis, but only if an anterior blastema was formed at the anterior wound of the latter was the pharynx then regenerated by the blastema.

Ishikawa and Ishii (1961) mainly confirm Kido's findings concerning regeneration of the pharynx in *Dendrocoelum lacteum*. In addition they discussed some biochemical factors mentioned later in Chapter 18.

Teshirogi (1962) studied morphallaxis in *Bdellocephala brunnea* and found de-differentiation in the gastroderm near the wound. Neoblasts were the principal cells forming the blastema, but he maintained that perhaps dedifferentiated gastroderm cells and fixed parenchymal cells also took part. The migration of cells was related to the rate and amount of regeneration. He found mitoses most frequent near the wound especially just over and parallel with the ventral nerve cords. He held that neoblasts regenerate the epidermis, nerves, muscles, testes, ovaries and probably also the copulatory organs, whereas the intestine was regenerated both from neoblasts and intestinal cells. He maintained that a sequence of tissue and organ formation took place "at its original level in the entire worm, e.g. pharynx regeneration is faster in level 2 than in level 1, and in level 2 faster than eye formation".

McWhinnie and Gleason (1957) found neoblasts the only cells to undergo mitosis, with a peak rate on the 3rd day. Oriented migration of neoblasts started on the 2nd day after wound formation and continued for 4–6 days thereafter. The neoblasts were mainly those which replaced lost parts. Colchicine restricted regeneration.

Fedecka-Bruner (1960) stated that regeneration of the copulatory organs occurred later than that of all other organs, and that no regeneration occurred if both head and pharyngeal regions were removed simultaneously. This is of considerable interest and will be discussed in Chapter 23.

Teshirogi and Yamada (1960) studied the regeneration of the copulatory organs in *Bdellocephala brunnea*. At temperatures between 17° and 34°C the frequency of regeneration of copulatory organs was greatest in postpharyngeal pieces, where ovaries were never regenerated (just as heads in this genus). Regeneration of the copulatory organs was infrequent just in front of the pharynx, and it never occurred in the forepart of the body, but at temperatures between 0° and 14°C copulatory organs were regenerated in this region also. This holds for starving animals, but if the worms were fed, copulatory organs were regenerated in all levels regardless of temperature, although the frequency was always highest in postpharyngeal pieces. This, the authors hold, "is the reverse reaction to the results of the head-frequency in *Bdellocephala brunnea,* and in accordance with the tail-frequency in *Bdellocephala punctata* as shown by Brøndsted" (Fig. 20).

Bandi (1959) held that cerebral ganglia were not formed by neoblasts but by the proliferation of the ventral nerve cords.

Gazco (1958) removed the pharynx and the pharyngeal sac from the ventral side. The pharynx regenerated after 8 days in *Dugesia lugrubris,* but in *Dendrocoelum* only after 14 days. According to Gazco regeneration of the pharynx started anteriorly in the empty cavity and grew posteriorly. He supposed that the regeneration began with de-differentiation of the old tissue cells; his pictures, however, suggest that neoblasts performed the regeneration. He found mitotic activity with an increase in nuclear volume 3–4 times, and a loosening of the chromatin ("se relache"). Gazco's pictures are indistinct and not convincing.

It will be seen from this short review that the problem of morphallactic mechanisms are still confused, even in recent literature, and we still have a long way to go before they are solved satisfactorily. In addition the mechanisms may differ in time and procedure from segment to segment in the same species, and also in some degree from species to species, although I am convinced that the basic principles are the same.

It may be mentioned that in Rhabdocoeles, Hein (1928) ascribed the regenerative processes exclusively to cells designated as neoblasts in modern terminology, with the exception of the epithelial cells, which together with neoblasts regenerate new epithelium. Levetzow (1939) found in Polyclads that cellular regenerative material was uniform, consisting of (1) mesenchymal cells and (2) formative cells (= neoblasts?); morphallaxis was said not to occur.

Pechlander (1957) studied regeneration in the alloeocoel *Otemesostoma* in which only transverse cuts are of value experimentally. Neoblasts closed the wound and only tissues of gland cells, muscles, sense organs, yolk follicles, intestine and partly amputated pharynges were regenerated.

In conclusion it should be mentioned that Abeloos (1965) has dealt with morphallaxis in Annelids in a valuable paper.

CHAPTER 9

INDUCTION AND ORGANIZATION

THE basic experiments of Spemann and Mangold (1924) and of Spemann's collaborators and pupils concerning "organization" and "Organisatorstoffe" in the 1920's and 1930's on embryological processes stimulated many workers to investigate the same principles in planarian regeneration. It is only fair to mention that Child (1914) had already said that the building of a new head was not only the restitution of a missing part but the first step in the development of a new individual and (1929 a, b), discussing his gradient hypothesis, he emphasized that the new head was the dominant organizer; we here encounter an idea similar to Spemann's "oberlippe" as organizer in the amphibian gastrula. Several of Child's co-workers and pupils, e.g. Watanabe (1935a), also held that the head was an inductor and self-determining.

Two points must be considered separately: (1) if a head is cut away behind the head-ganglia, a new animal is formed from the ensuing anterior blastema giving the impression that the head is in fact dominating and organizing the regeneration process leading to the formation of the new animal; (2) if a head is transplanted elsewhere in the body of an entire or decapitated animal, will the implanted head then remould the worm according to these organizing forces? There is debate about this, so let us therefore look into the controversial literature concerning point (2).

Santos (1929, 1931), using *Dugesia dorotocephala* and *D. maculata (= tigrina)*, cut holes in the host into which head-ganglia, bearing parts of heads were implanted; if the implants were made prepharyngeally they acted as "organizers"; implanted postpharyngeally they sometimes reversed the polarity of the host and inhibited head regeneration in the decapitated host. The implant could determine development of post-cephalic outgrowths or further reorganization in the host body, with the development of a new pharynx and postpharyngeal region. At times many curious protuberances ensued. Sometimes the grafts migrated or were resorbed and sometimes their polarity was impossible to recognize. The induction was not species-specific: implants of *Dugesia dorotocephala* sometimes induced in *D. maculata* hosts. Implants of heads adjacent to the head of the host were resorbed. Santos concludes that the head was the most principal part of the body both in activation and in dominance.

Both these species also propagate by fission in the postpharyngeal part of the body. As we shall see in Chapter 16 this part of the body in fission-propagating species is a region where neoblasts are easily activated. I therefore think that this region of the body in such species is not well suited for Santos' transplantation experiments, because clear and unambiguous results cannot be achieved here because several unexpected and unforeseen strategies of "induction" and "organization" may happen here.

Okada and Sugino (1934, 1937) in *Dorotocephala gonocephala*, in contrast to Santos but without precisely defining their technique, found that the orienting induction of the transplant was unaltered, nor did it reverse the polarity of the host, but the transplant (head often without ganglia) induced pharynx formation in the host, the new pharynx being made of tissue from both transplant and host. Okada (1941) broadened the experiments, by implanting a prepharyngeal piece into the pharyngeal region of *Dugesia gonocephala*. A pharynx was induced not only in the host but also in the implant. (We shall see later that there seems to exist a "pharyngeal zone around the pharynx which is able to induce the formation of a pharynx). Okada found that in the anterior part of the implantation area ciliary movement was reversed. In another experiment Sugino transplanted a postpharyngeal segment in front of a prepharyngeal segment, both with normal polarity; a head regenerated from the anterior surface of the postpharyngeal implant and a tail from the posterior surface of the prepharyngeal segment. In some of his experiments a reversal of polarity followed. Sugino deduced that the head had no influence on the polarity but the new tissue had its polarity determined by a pre-existing gradient from a higher to a lower threshold.

Rand and Brown (1926) implanted heads in a decapitated host to find out if the graft was able to organize and so inhibit the formation of a new head from the decapitated host, but a new head from the host was regenerated even though the blastemata were removed several times. If the graft was implanted close to the decapitation area the graft developed into the head of the host, but as expected no reversal of polarity occurred.

Spirito (1935) in *Dugesia torva* found that foreparts implanted into the genital tract did not reverse the polarity, but supplementary pharyngeal structures were found.

It seems to me that the differentiated adult head has not the capacity to organize further an already organized adult body segment; this can only be done by a blastema (after decapitation) if such is formed after implantation. The organizing and inducing factors seem only to act when embryonic blastema cells or neoblasts are present in great masses as a result of the implantation procedure. I think that the conflicting results obtained by various authors were due mainly to the varying conditions following conflicting and "confusing orders" which the neoblasts received on account of the crude transplantation techniques used.

It therefore seemed advisable to employ a transplantation technique omitting blastema formation as far as possible. Brøndsted (1939) invented a simple transplantation technique which has proved very effective in many varied experiments. The animals were anesthetized for about 1 hour in 1/5000 to 1/20,000 nicotine sulphate, varied according to the species and the aim of the experiment. The desired segments were cut cleanly from animals previously placed on paraffin wax, which had been lightly flamed. The pieces to be transplanted were placed on glass rings covered with "millersilk" ("Müller gace", a technique devised by Shotté), and arranged in the desired position with the clean wounds matching one another as exactly as possible. Nicotine sulphate was carefully pipetted over the animals, the rings placed in Petridishes containing nicotine water just level with the upper rim of the glass rings, but not overflowing the silk bearing the transplants. The dishes were placed in the dark preferably at temperatures of 5–10°C. Pipetting about every hour was necessary to remove exudates from the wounds. After about 24 hours the cut edges adhered so strongly that the nicotine solution could be replaced cautiously by tap-water. After 48 hours the transplants generally adhered sufficiently firmly for one to pipette gently enough water into the Petri-dishes to swamp the silk, and for the transplants then to swim and sink to the bottom. The water was changed and after 1 or 2 days the chimeras could be handled normally.

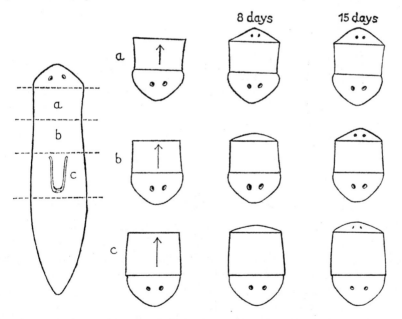

FIG. 68. *Dugesia lugubris*. Arrows indicate polarity anteriorly. Explanation in text. (From Brøndsted, 1939.)

This method was used to see if the adult head really could reorganize adult segments. Figure 68 shows one set of experiments: a head (homotransplant) was grafted on to the posterior surface of segments *a, b* or *c*. In no case did the adult segment respond to any "organizing" influence from the graft, i.e. no reversal of polarity took place. From Fig. 68 it can be seen that the eyes appeared later in the blastemata from posterior segments in agreement with the time-grading seen in *Dugesia lugubris* (Chapter 5). When the segments attained a fair regeneration level chimeras were formed consisting of two animals dragging in opposite directions. According to Child's gradient hypothesis the transplanted head should have had a good opportunity to "organize", i.e. reverse the polarity of the segment in which the head had been transplanted, at least the pharyngeal ones, the "weaker" in Child's "morphogenetic gradient".

To strengthen these results some other experiments were done. Figure 69 shows the procedure. When eyes were discernible in the anterior blastema

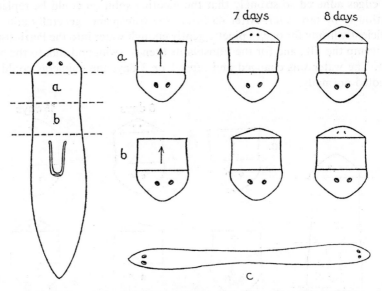

FIG. 69. *Dugesia lugubris*. Explanation in text. (From Brøndsted, 1939.)

of the segment, it was cut away in order to see if the transplanted head was then able to "organize" the morphology of the segment; this did not happen either with the transverse segment *a* or the more caudal one *b*. It might be argued in explanation that when a new head had already regenerated from the segment, however small, it could not be over-run by some "organizing" influence from the transplanted head. Figure 70 shows that even when a barely formed blastema was cut away, the result remained the same.

As stated in Chapter 10 a transverse piece sometimes contracted so severely that no blastema formed. In such an instance one might expect that the transplanted head would then have the opportunity to "organize". But despite

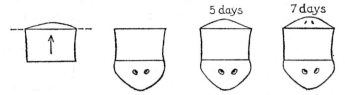

FIG. 70. *Dugesia lugubris*. Explanation in text. (From Brøndsted, 1939.)

this favourable situation a head regenerated from the anterior surface of the segment when a cut separated the coalesced parts, even if the rate of head formation under these circumstances was much delayed (Fig. 71).

The possible "organizing" influence of the transplanted head in reversing the dorso–ventral axis was investigated: a head was transplanted upside down in the anterior surface of a segment. After several days the segment had regenerated its own head in accordance with its own dorso–ventrality as seen in Fig. 72 where the grafted head is drawn without eyes because only the ventral surface is shown.

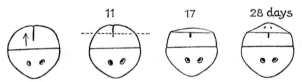

FIG. 71. *Dugesia lugubris*. Explanation in text. (From Brøndsted, 1939.)

FIG. 72. *Dugesia lugubris*. Explanation in text. (From Brøndsted, 1939.)

In several places it has been pointed out that *Bdellocephala punctata* (Pallas) is not able to regenerate a head from levels just anterior to the pharynx. Figure 73 (see also Fig. 20) shows incisions at various levels: regenerative power was not lost in the posterior parts of the body; on the contrary the more posterior the cut the faster the posterior regeneration of a tail. Figure 73 also shows that in the segment containing the copulatory organs regeneration followed that of the main body axis. Brøndsted (1939) suggested the possible existence of two opposing regeneration gradients, analogous perhaps to the animal-vegetative gradients found in the egg of the sea-urchin (Runnström, 1928).

Using *Bdellocephala* Brøndsted (1939) suggested the idea that when a segment of the forepart was able to regenerate a head, some head-inducting substance might be released from the head (proven many years later by Wolff

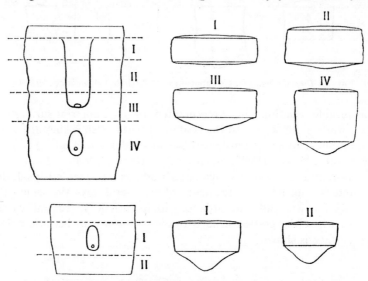

FIG. 73. *Bdellocephala punctata.* Cuttings at various levels in the midpart of the body. Explanation in text. (From Brøndsted, 1939.)

and Lender (1950a, b) and Lender (1950, 1951a,b,c) and discussed later in this chapter). This might then diffuse into segments from the posterior part of the body inducing head formation in them. Figure 74 shows the experiment: *a* was grafted to *b* with reversed polarity; after healing three groups of chimeras were cut as designated in A, B and C. In none of these did head formation occur, except where the edge of C was sufficiently thick, but here as seen in Fig. 75, it is separated from the tail piece by an unpigmented zone indicating that the blastema of this piece was without head-forming qualities. The interesting point is that such a narrow rim was able to reverse its polarity, as in Kido's experiments referred to above.

It seems to me that all these experiments contradict the idea of the dominating and organizing power of the head of the adult planarian. The splendid experiments of Dubois and Kolmayer (1959) showed that neoblasts from the posterior body of *Dendrocoelum* (which is unable to regenerate a head from this region as in *Bdellocephala*) arrived at the anterior portion of the body and there took part in the regeneration of a head. It may be deduced, therefore, with a fair degree of certainty that the posterior part of the body in these two species simply lack the potentially to respond to head inducing substances, a conclusion already drawn by Brøndsted (1939). It remains, however, to decide with certainty that time plays no role in this matter:

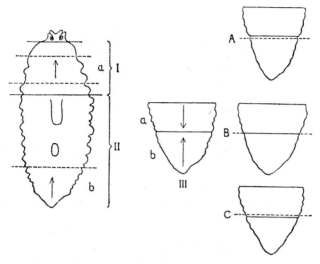

FIG. 74. *Bdellocephala punctata*. Explanation in text. (From Brøndsted, 1939.)

experiments should be conducted to find out if heads can be regenerated from the posterior parts of these species after a prolonged interval.

It should be mentioned that neoblasts migrating from the posterior part of these species may regenerate heads in the forepart of the animals, disproving the notion held by Goetsch (1932) that head- and tail-determining cells exist in the planarian body. The same author discussed the "organizer" problem in planarians (1926, 1929).

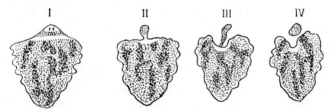

FIG. 75. *Bdellocephala punctata*. Explanation in text. (From Brøndsted, 1939.)

If an anterior segment of considerable width is transplanted with polarity reversed into the anterior surface of a posterior piece (Fig. 76), then the head-forming influences in the anterior piece are blocked and the posterior segment regenerates a tail. But narrow grafts are often discarded as though there were a serological difference between pieces with opposite polarity, even from the same animal.

To study the possibility of the presence of diffusible head-inducing substances, sterilized 3 % agar blocks were laid firmly on to the anterior surface

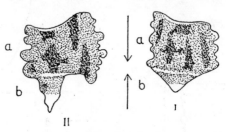

FIG. 76. *Bdellocephala punctata*. Explanation in text. (From Brøndsted, 1939.)

of anterior pieces for some hours; the agar blocks were then removed and fixed to the anterior surface of posterior segments to see if head-inducing substances had been taken up in the agar as they would then be released to the posterior segments. On removal of the blocks 2 days later, blastemas had formed, but no heads. The result of this rather crude experiment may not be very convincing, but taken together with all the other experiments cited above, I think it may be inferred that Huxley and de Beer's (1934) suggestion that the planarian head is equivalent to the "organizer" of the amphibian embryo is incorrect.

Brøndsted (1939) summarized the situation as follows:

> Organizing capacities in the adult planarian body are present only in a masked or latent manner. They can unfold themselves only when a blastema is formed. ... The blastema is, of course, made of cells from the grown-up body itself, but only when the cells in the blastema have been together for some unknown time are they made sensitive to influences from the adult body, and then according to these influences determined in accordance with the polarity and capacities of the body. But then the blastema takes the lead, and besides differentiating itself, gives off organizing influences to the body, so strongly marked that they are able to remould the adult tissues (morphallaxis).
>
> Therefore, we must look for competent tissue (Waddington), and for tissue containing organizing forces comparable with the organizers in embryonic development only in the blastema. Adult tissue on the contrary is only competent under the influence of forces generated in the blastema. ... Further, and very strong, evidence that the formation of a blastema of a craniad facing wound is necessary for the rebuilding of the whole organism is afforded by the fact that when a wound is closed up, so that no blastema is formed, no morphallaxis takes place. The forces necessary for morphallaxis therefore must be generated in the blastema.
>
> All this suggests that in the fully differentiated adult organism we have to deal with inhibiting factors controlling the interaction of the various organs and cells. In the regenerating organism "organizing" forces control the building up of the animal until a new level of interaction between the organs has been established.
>
> If this idea be accepted, it is easily understood that the differentiated adult tissues do

not exercise any organizing influence on other adult tissue, as has now been shown for the head of *Planaria lugubris* and body-segments of *Bdellocephala punctata*.

It seems to me that although in somewhat old phrasing these citations are still valid and their correctness is borne out by many later experiments.

In the older literature we sometimes meet strange ideas as to the determination of differentiation during the regenerative processes. Steinmann (1910) proposed the hypothesis that differentiation of organs distant from the wound was a sign of determination by "die Gesamtheit der Zellen des Regeneranten"; and (1926) a kind of metaphysical way of thinking was put forward in a plan regulating restitution processes that a sort of "eine Über-planarie" might exist, a truly vitalistic idea.

Li (1925) vaguely hypothesized that regenerative forces were comparable to radiating energy, and Keil (1924) in *Polycelis nigra* stated that the new eyes in the blastema started their formation close to the old tissue, their position determined by unknown forces. Lus (1926), using marine triclads, produced heteromorphoses and found that "Kopfreize" determined the blastema, as did all other stimuli in the body. Other authors, e.g. Gebhardt (1920), Abeloos (1930), Kahl (1936) and Morgan (1902), thought that the old tissues determined the blastemata, the regenerative material being toti-potent, but determined by the regenerant. Although the older literature was confusing, many workers were seeking to do what Wolff and his colleagues subsequently achieved.

A little anecdote pertinent to the confusion prevalent in the older literature may be narrated here. In the 1930's when I started my planarian experimen-tal work in Runnström's laboratory as a guest, the late Professor Peterfi visited us in Stockholm. When I told him about my plans, he said, "Don't start this work, you know that the planarians can do all sorts of strange tricks." This remark of course further stimulated my curiosity and endea-vours. Time has shown that an ever-increasing group of scientists have taken up the challenge, and even if much has still to be unravelled, much has since been achieved.

The outstanding work of the Wolffian school has been admirably reviewed by Wolff and Lender (1962) and by Lender (1963). In this chapter we are concerned with induction processes, and it was in this field that Wolff and his collaborators brought substantial progress. In Chapter 7, neoblasts were discussed at some length and here we shall discuss how neoblasts might be induced to differentiate.

An old and all-important question in planarian regeneration—as in all regeneration—is this: what role do the adult tissues play in the organization of the neoblasts assembled at the wound? Is the blastema itself the decisive factor in determining what regenerates?

Figure 77 shows an experiment (Brøndsted, 1942a), where a segment was cut out of the animal, the two remaining surfaces were then cut cleanly, so that no protuberances could cloud the result. Regeneration took place in such a way that the old parts remained intact, only furnishing material to the new piece forming between them. This was fairly good evidence for regarding adult tissues as the decisive factor in regenerative processes—at least in this experiment. But other experiments seem to indicate either that the blastema directs regeneration or there is a co-operation between inhibitory forces both from adult tissues and the blastema. Here let it be said that I think that adult tissues give their inhibitory information to the blastema, which, after having been determined, release the necessary information to the adult tissues, if these have to reorganize the whole body— as in most morphallaxes.

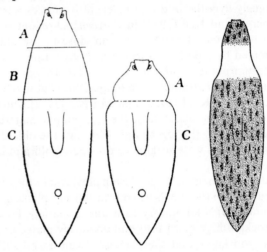

Fig. 77. *Bdellocephala punctata.* Explanation in text. (From Brøndsted, 1942.)

In this chapter we have already seen that the idea of the adult head as the all-important inductive and organizing agent, as held by Child and some co-workers, cannot be true. Figure 78, 1–6 shows how the Wolffian school looks at this cardinal question. In figure 1 the question mark meant that the authors thought the blastema formed after decapitation was determined by the base, i.e. adult tissue, but they were not sure. In Chapter 7 we discussed fine nerve fibres from the lateral nerve cords might penetrate between the neoblasts in the blastema and induce brain regeneration. It is still an open question as to whether the regenerating brain may be differentiated from the neoblasts without the co-operation of existing nerve fibres. Sengel (1960), in an elegant experiment, cut off the anterior blastema formed after decapitation in *Dugesia lugubris*; the blastema was cut away after a delay of 2–3 days, but even though histological investigation showed that the blastema seemed to consist of undifferentiated neoblasts, the possibility that nerve fibres had already penetrated the blastema from severed nerve cords cannot be excluded. The blastema was cultured in a standard solution (a modified Holtfreter solution, Wolff and Haffen, 1952). The anterior blastemata developed heads, the posterior ones tails, but none regenerated pharynx or intestine when grown in

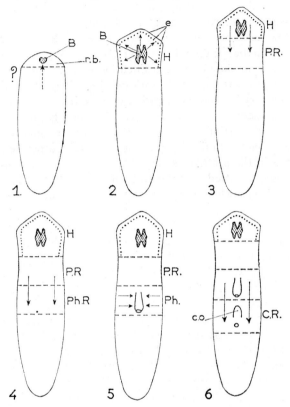

FIG. 78. Diagram interpreting the process of induction. 1, Determination of the blastema from the body. ?, Induction of the brain from the old parts. 2, Induction of eyes from the brain. 3, Induction of prepharyngeal region from the head. 4, Induction of the pharyngeal zone from the prepharyngeal one. 5, Auto differentiation of pharynx or from the neighbouring tissues. 6, Probable induction of the copulatory organ from anterior regions. (From Wolff and Lender, 1962.)

isolation, meaning that blastemata removed after 2–3 days were already determined, although not yet differentiated. If the anterior and posterior blastemata were united by their cut surfaces, they regenerated both pharynx and intestines. I think that the question mark on Fig. 78, 1 may be disregarded as these experiments support the idea that the blastemata were determined in the first 1–3 days from the adult tissues, namely, before they were separated from the base.

Figure 78, 2 shows the crux of a series of brilliant experiments by Wolff and Lender (1950a,b) and by Lender (1950, 1951a,b,c). These authors found in *Polycelis nigra* (a species characterized by a marginal string of eyes in the anterior third of the body), that when the brain was excised, it would regenerate in 3–4 days and the eyes about 7 days after their removal. But if the brain was removed, even if repeatedly during regeneration, no eyes were formed. Likewise, if the brain was heavily X-irradiated (with the eyes screened) and then the eye-containing rim removed, no eyes were regenerated. If the neoblasts near the brain were killed by mild X-irradiation, neoblasts from nearby regions regenerated removed eyes. They concluded that the brain was necessary for the regeneration of eyes, in other words, the brain induced eye-regeneration. After Lender (1956b) had grafted a piece of the eye-bearing lateral portion of the animal caudally presumably out of reach of the influence of the brain no eyes were regenerated before a new brain had developed in the graft. It is very interesting that if a non-eye-bearing part of the body is grafted near the brain of a host, no eyes are regenerated in this part and only when a brain is regenerated in the graft are eyes also regenerated.

Lender (1951c) found that the brain of *Dugesia lugubris* was able to induce eyes in *Polycelis*. The inductive agent was therefore not species-specific.

These important results were followed up by penetrating experiments by Lender and co-workers. Lender (1955) named the hypothetical induction substance *organisin*. He made a brei of the heads of *Polycelis*, *Dugesia lugubris*, *D. gonocephala* and *Dendrocoelum lacteum*. The breis induced eye-formation in *Polycelis*, but the brei from *Dendrocoelum* only feebly. Neither 70% ethanol nor desiccation destroyed the effect. Centrifugation showed that the inducing factor was found mainly in the supernatant fluid, showing that it was water-soluble. Lender (1956a) found that the inductive force depended on the concentration of organisin in the solution, and that this declined posteriorly, a brei from the caudal region inducing only feeble eye-formation. He suggested that organisin may be immobilized in the cells of the normal caudal parts of the animal. Lender (1956b,c) showed that brain organisin inhibited regeneration of a new brain, although eyes could still be regenerated. In contrast to the inducing substance, the inhibiting substance was specific.

Figure 78, 3 and 4 show schematically experiments by Lender (1956c) that the head and prepharyngeal region induced a "pharyngeal zone", which itself induced differentiation of a pharynx. It is, however, not clear if pharynx formation was really autodifferentiated. Sengel (1951, 1953) found that an anterior blastema, regenerating a head, induced a zone called "pre-pharyngleane", in which the pharynx itself regenerated.

Kido (1957) also studied the induction of a pharynx in *Dugesia dorotocephala* arising from the old ideas of Santos (1929, 1931), Steinmann (1933), Okada and Sugino (1934, 1937) and Miller (1938) that the head may induce a pharynx in the posterior part of the animal. He tried to see if the head might induce—in a manner similar to the devitalized "Oberlippe" in amphibian gastrulae. He fixed the heads in 96% alcohol and then placed them in water at 60°C. After making a brei in agar he implanted the centrifuged extract into the hind part of the animals, but no pharynx induction occurred. Kido thought that it was the living brain which induced in the older experiments because he found induction in two out of eight instances with living heads. In the two positive inductions the cerebral ganglia had connected with the nerve cords of the host, but in the six negative ones no connections were found. This finding of Kido was a fairly good indication that living nerve tissue was the inductive agent.

Lender (1960) stressed the notion that the inducing and inhibiting substances were humoral and not necessarily bound to cells. He also found that the soluble inducing substances followed Child's gradient. It must be emphasized here that this hypothesis does not in the least strengthen Child's hypothesis of a metabolic gradient due to a respiratory gradient.

Figure 78, 6 shows that the Wolffian school thought that induction from the pharynx-zone was necessary for regeneration of copulatory organs. The diagrams suggest that a chain of inductions, starting from the head blastema, organize the whole body.

Lender and Deutsch (1960) found that a brei from chick embryos also induced eye-formation in *Polycelis nigra*. Here the "organisin" seemed to be in the sediment and not the supernatant fluid. This result was surprising, because it showed induction in invertebrates by substances found in vertebrates, and because the substances sedimented differently in the centrifugate. This may suggest that non-specific substances may induce specific organ formation. It reminds one of some analogies, i.e. the non-specific agencies causing artificial parthenogenesis, giving rise to doubt about specificity in the case of "organisin". The situation also vaguely reminds me of the "induction" problem of the amphibian "Oberlippe", where numerous investigations have shown that non-specific substances, including inorganic ions, also induce. I think that in all these instances, as well as in planarian regeneration, there are problems not yet understood, all involving a decisive time-factor. We shall discuss these problems in more detail in the concluding Chapter 23.

Stéphan-Dubois and Lender (1956) discussed the possible existence of "wound hormones". They found that neither burning, electro-coagulation nor irradiation elicited migration of neoblasts; only real wounding had this power. They suggested that perhaps "wound hormones" were neurohormones (see Chaper 13), because no eye regeneration occurred when the brain and the eye-bearing region were removed. The authors could not find nerve fibres reaching the eye-forming region before the eyes regenerated, and suggested that the neoblasts near the eye region induced, and not the neoblasts near the brain. From this they deduced that eye induction was of humoral character.

In addition they made a brei, which was centrifuged thoroughly; the super-natant induced and was not species-specific. Seventy per cent alcohol and heating to 60°C did not remove the inductive capacity. The inducing sub-stances were found in concentrations declining from head to tail, following Child's gradient. The "organisin" concentration in heads was four times that in tails. The authors thought that the "organisin" was confined to the reserve cells in the tail, but liberated by the brei process.

I must emphasize that these induction phenomena only concern eye formation, and cannot therefore be taken as a proof that regeneration in general follows the gradient. As said before (p. 31) tail regeneration has its own gradient opposite to that of Child's.

Pentz and Seilern-Aspang (1961) confirmed Lender's results that the brain induced eye formation in *Polycelis nigra*. They found it likely that a radially shaped area of inhibition developed around every regenerating eye, a pheno-menon responsible for normal eye regeneration. The authors suggested that

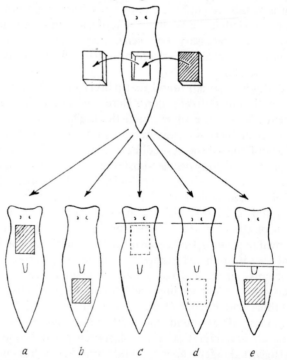

| a | b | c | d | e |

FIG. 79. *Dendrocoelum lacteum*. Diagram: a, copulatory organs are transplanted to the forepart of the animal, this part is able to regenerate head; b, the graft is transplanted to the hind part, which is not able to regenerate a head; e, the same, but the forepart is removed. In these three instances the graft healed in; c, graft transplanted to forepart, d, to hindpart; in both cases the animals are decapitated, and in both cases the grafts are resorbed. (From Ortner and Seilern-Aspang, 1962.)

the eye pattern in *Polycelis* was the result of an antagonism between the inductive force of the brain and the inhibiting areas.

This idea is of great interest because it fits in with the general conception of an essential equilibrium between inductive and inhibitory forces in all morphogenesis, reminding one of Wigglesworth's (1940) experiments on the regeneration of bristles in *Rhodnia*.

Seilern-Aspang (1957, 1958, 1960a) described *in vitro* cultures of the contents of the cocoons of *Dugesia torva*. His experiments are interesting and suggestive, although I find their interpretations very hypothetical. If elaborated with histochemical methods they might be of value in regeneration research. The author drew interesting parallels with the free amoebocytes of the Acrasiae (*Dictiostelium*, etc.) which produced interacting substances marshalling migration, resulting in a building up of the multicellular organism.

Ortner and Seilern-Aspang (1962), in *Dendrocoelum lacteum*, published several interesting experiments concerning an "Organisationsfaktor" (Fig. 79) with somewhat theoretical interpretations, summarized thus:

> Es konnte dabei gefunden werden, daß der Organisationsfaktor bei einem Gleichgewicht des Systems inaktiv ist. Der Organisationsfaktor läßt sich auch nicht aktivieren durch rein qualitative Störungen des biologischen Systems, wie etwa durch ortsfremde Austauschtransplantate und durch Zentrifugieren. Auch Wundsetzung ohne Gewebsentnahme ist unwirksam.
>
> Dagegen gelingt eine Aktivierung des Organisationsfaktors durch Störungen des quantitativen Gleichgewichtes. Sowohl Gewebsentnahme als auch Gewebszunahme lösen die Tätigkeit des Kontrollsystems aus: auf Gewebsentnahme reagiert der Organismus mit Regeneration, auf Zugabe von Gewebsüberschüssen mit deren Unterdrückung. Bei gleichzeitiger Gewebszugabe und Gewebsentnahme fällt diese unterdrückende Wirkung des Organisationsfaktors auf das eingeschobene Gewebe aus. Zugabe und Entnahme kleinster Gewebestücken lösen nicht den Organisationsfaktor aus, für den also eine gewisse Toleranzbreite angenommen werden muß. Röntgenversuche zeigten, daß nach Gewebsentnahme auch ohne einen Regenerationsvorgang der Organisationsfaktor aktiviert wird.
>
> Der isolierte, nicht regenerationsfähige Hinterkörper von *Dendrocoelum lacteum* vermag nicht, den Organisationsfaktor zu aktivieren.

It must, however, be emphasized that the authors neglected the fact that this species is able to regenerate a tail from the hind part.

It should also be mentioned that Török (1958), experimenting with *Dugesia lugubris*, held that the brain regenerated from the old nerve trunk, and that the brain was necessary for eye regeneration, in agreement with Lender. He was of the opinion that the time-graded regeneration field was in some way connected with the number of nerve cells. If this were the case it would be hard to say why no head ganglia and eyes regenerate from the parts behind the pharynx in *Dendrocoelum* and *Bdellocephala*, where tail regeneration readily follows transverse cuts.

4a PR

Vannini (1965) and his group in Bologna investigated induction of the gonads. In *Dugesia lugubris* the forepart of the body, containing the brain, ovaries and some testes, was cut away (Fig. 80). The testes in the hind part regressed completely, but regenerated again together with the ovaries after a head with a brain and eyes had reformed. In *Polycelis nigra* it was necessary

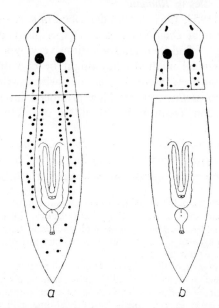

a   *b*

Fig. 80. *Dugesia lugubris*. Explanation in text. (From Vannini, p. 165, fig. 7 in *Regeneration in Animals*, ed. Kiortsis and Trampusch, North-Holland Publishing Co., Amsterdam.)

to remove the whole eye-bearing region to obtain regression of the testes (Fig. 81). In the discussion Vannini pointed out that if head regeneration was suppressed in the hind part by sympamine the testes never regenerated, but that if the head was cut away in front of the ovaries, regression of the testes did not take place. Lender in discussion (p. 175) said that neurosecretion did not seem to play a role in regeneration, but probably took part in the differentiation of the genital organs.

Ghirardelli (1965) also reviewed work along these lines and gave further evidence of the controlling action of the head for the development of gonads. If *Dugesia* or *Polycelis* were decapitated and split medially from the head end to a point just anterior to the pharynx, then a two-headed monster appeared, a well-known experiment. In *Dugesia*, if the one head was then cut away, the testes in the corresponding half-animal regressed (Fig. 82), but in *Polycelis*, where the median cut has to be deeper in order to remove

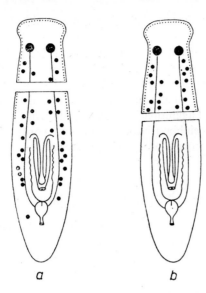

FIG. 81. *Polycelis nigra*. Explanation in text. (From Vannini, p. 166, fig. 9a, b, in *Regeneration in Animals*, ed. Kiortsis and Trampusch, North-Holland Publishing Co., Amsterdam.)

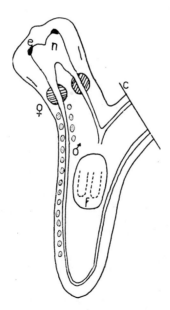

FIG. 82. *Dugesia lugubris*. Explanation in text. (From Ghiradelli, p. 180, fig. 3, in *Regeneration in Animals*, ed. Kiortsis and Trampusch, North-Holland Publishing Co., Amsterdam.)

all the eyes, numerous testes developed along the regenerated "median" nerve cords.

Fedecka Bruner (1964, 1965) irradiated *Dugesia lugubris* from the dorsal side where the testes are situated. The irradiation was given in doses only sufficient to reach the dorsal part, so the neoblasts accumulating along the ventral nerve cords were unaffected. After treatment the testes were destroyed, showing pycnosis and agglutination. After 15 days all debris was eliminated. Neoblasts from the neighbourhood of the nerve cords migrated dorsally to form the testes, the process progressing posteriorly.

CHAPTER 10

INHIBITION PHENOMENA

IN ORGANISMS such as planarians with a great capacity for regeneration one may well ask how individuality is preserved when equilibrium is disturbed. There is no doubt that in certain experimental and natural conditions, the individuality may be disturbed producing monsters and other heteromorphoses (Chapter 14). In most cases, however, wounding or removal of parts of the body calls forth mechanisms by which inhibition of superfluous organ formation is secured. These inhibitory mechanisms are, of course, just as vital as induction mechanisms in attaining a harmonious individual. Inhibition is therefore a basic problem for investigation in all morphogenesis. In planarians we have excellent subjects for studying the problem of inhibition, but again the literature is contradictory.

In order to obtain regeneration of head or tail when a piece of the body is cut away, it has been repeatedly emphasized in the literature that the wound should be kept open; Stevens and Borin (1905), Silber and Hamburger (1939), Brøndsted (1939), Dubois (1949) and many others have observed this phenomenon. A puzzling observation which most investigators have made is that

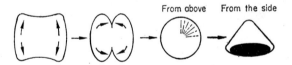

FIG. 83. Explanation in text. (Original.)

sometimes in short transverse sections the muscular activity in trying to diminish the wound surface was so strong that the left and right side of the body near the wound met, completely closing the wound (Fig. 83). No regeneration followed though often the segment after some days formed a curious hood like a short hollow cone. How may this failure to regenerate be explained? In the first place Dubois (1949) showed conclusively that an open wound elicited a migration of neoblasts to the wound forming a blastema. It seems plausible that some neoblasts had time to migrate to the wound before it was completely closed and absence of neoblasts cannot therefore be

101

the only cause for non-regeneration so there must be others. The arrows on Fig. 83 indicate cephaled or caudad polarity; when the wound is completely closed the two opposing polarities met one another, perhaps suggesting that inhibitory mechanisms inhibit both halves from regenerating the same thing, a head or a tail, but when a cut is made as in Fig. 84 a head or tail is promptly

FIG. 84. Explanation in text. (Original.)

regenerated provided that the wound does not close again (Brøndsted, 1939), because then the polarity can again have play.

Figure 85 A, B, C (Brøndsted, 1942) shows an experiment in which the animal was decapitated and the tail transplanted into the anterior wound blocking head regeneration. Even if the head were trausplanted to a wound at the side just caudad to the transplanted tail, this head (the animal's own!) would follow its own course in moving forward, with the decapitated animal trying to go forward with its useless tail hanging on, so the chimera would travel in a circular course determined by the body and the now alien head.

It is not enough to say, as held by Schultz (1902b), that mechanical hindrance for regeneration is built up by the closure of the wound. Figure 77 shows an experiment where the sector *b* was cut out and discarded and the

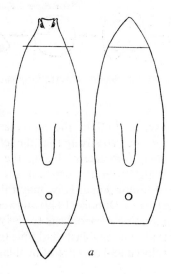

a

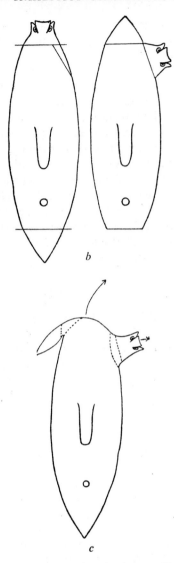

FIG. 85. *Bdellocephala punctata*. Explanation in text. (From Brøndsted, 1942.)

a-piece transplanted to the *c*-piece with the same polarity. Here a "mechanical hindrance" should surely have prevented regeneration, but the lost tissues and organs nevertheless regenerated in between. The experiment, of course, disproved the claim, also held formerly by myself, that an open wound was necessary for regeneration. The experiment has not yet been investigated histologically, which would possibly provide important information relevant to morphallactic processes and the role neoblasts play in regenerative processes not involving open wounds. It may also have an important bearing on the basic regenerative processes, discussed in the concluding Chapter 23.

Let us again look at experiments bearing more directly on the problem of inhibition. Rand and Ellis (1926) made planarians regenerate two tails by splitting the animals medially from the tail tip towards the pharynx; preventing coalescence of two halves led to each regenerating a whole tail; if one of the tails was partly cut away, the stump promptly regenerated a tail; but if one tail was cut away at the base, then a new tail did not regenerate. An analogous experiment may be performed by splitting foreparts. Rand and Ellis interpreted this as being due to inhibition: "... is specifically due to the presence, in appropriate proximity and relations, an *equivalent* of the structures which have been removed... There is demonstrated a dominance of one part of an organism over another." Roulon (1936b) found similar inhibitory forces from posterior cut surfaces.

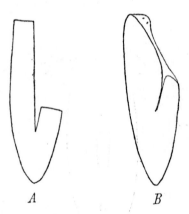

A B

FIG. 86. *Dugesia lugubris*. Explanation in text. (From Morgan, 1902a, after Brøndsted, 1946.)

An important example of inhibition—non-inhibition has already been given by Morgan (1902a) (Fig. 86) who wrote:

> If the anterior end is cut from the worm, the body partially split into two equal parts, and one of the halves is cut off near the bottom of the last cut, as indicated in fig. 14 (Morgan's paper), there are exposed two cut-ends at different levels. The pieces will fuse along the middle line unless they are for a day or two repeatedly separated. If the pieces fuse along the middle line a single head develops, and this head is always at the anterior cross-cut. New tissue appears along the inner edge of the longer half-piece that is at its most posterior end continuous with the new tissue covering the posterior cross-cut. A new head is not produced by the latter, and this

region is slowly changed into the side of the new worm. If, however, the longer and the shorter half-pieces are kept from fusing along the middle line a new head generally develops on the posterior cross-cut.

It is worthwhile to bear this experiment in mind when discussing the principles of inhibition.

Li (1928) conducted the following experiment: decapitated animals were cut along the median line, separating the forepart into two; each part regenerated its own head if they were held separated. If both regenerated foreparts were then cut away leaving two stumps, each again regenerated its own head, but if the cut was made more posteriorly so that only one common surface was exposed, then only one head was regenerated. The existence of two foreparts, in fact two individuals, did not impose duality in the hinder part of the body which was to be expected according to Child's notion of the head as predominant indicator.

The problem is why inhibitory forces display themselves in some instances and not in others. Morgan's experiment cited above seems to me to unravel some of the riddle: my idea is that the inhibitory forces (discussed later in this chapter) travel much faster through blastema tissue than through intact adult tissues. A few experiments supporting this idea may be presented.

Figure 87 (Brøndsted, 1942) shows *Bdellocephala* after decapitation deprived of the median part of the body down to the anterior level of the pharynx. It will be remembered that this species is unable to regenerate a head from this level. Each of the two "arms" regenerated a head, even if one "arm" was shortened. Perhaps even more striking was the experiment shown in Fig. 88 (Brøndsted, 1942) in which a "window" was cut away in the forepart of the body somewhat anterior to the pharynx; after some time a new little head developed out of the "window", the existing old head showing no inhibitory effect on the regeneration of a new head from the middle of the body. In order to see if a blastema on the anterior surface of a decapitated animal developed inhibitory substances, a "window" was cut out in the same manner as in the foregoing experiment, but there was no difference in the result. (See also this book, Fig. 23.)

True inhibition is, however, effected through the blastema, which seems proved by the following experiments (Brøndsted, 1954): when an animal was decapitated by a clean transverse cut only one head was regenerated from the ensuing blastema; but every part of the whole body was able to regenerate a head, and hence a whole animal, and would do so if that part was separated from the rest of the body (see Chapter 5). The experiments were carried out on *Dendrocoelum lacteum* and *Bdellocephala punctata*.

In *Dendrocoelum* decapitated transverse sections of the forepart of the body from a great many specimens were allowed to regenerate; the segments were divided into five equal groups. Figure 89 shows the course of the experiment. After 24 hours the lateral third of the segments in the first group were

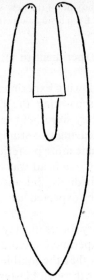

FIG. 87. *Bdellocephala punctata.* Explanation in text. (From Brøndsted, 1946.)

FIG. 88. *Bdellocephala punctata.* Explanation in text. (From Brøndsted, 1946.)

FIG. 89. *Dendrocoelum lacteum.* Explanation in text. (From Brøndsted, 1954.)

cut away and allowed to regenerate individually; after 48 hours the lateral thirds of the second group were cut away and allowed to regenerate; after 72, 96 and 120 hours the same was done with groups 3–5. Meanwhile the pieces containing the median parts of the blastemata had regenerated heads with eyes after about 150 hours at 20°C. Other experiments showed that isolated lateral thirds of segments removed from the same level lying laterally in the time-graded field, regenerated heads with eyes after about 170 hours. Figure 90 shows that the lateral thirds, separated after only 24 hours from

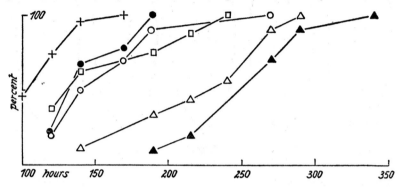

Fig. 90. *Dendrocoelum lacteum*. Further explanation in text. (From Brøndsted, 1954.)

the whole decapitated segment, regenerated heads and eyes at about the normal speed; separation after 48 and 72 hours allowed sufficient time for the development of appreciable inhibition of regeneration which was even more evident after separation at 96 and 120 hours. Thus the lateral thirds, when still part of whole segments containing the "high-point", can still regenerate heads when separated, but are hindered from doing so before separation from the "high-point". The most plausible interpretation of these experiments was that some inhibitory influence from the "high-point" reached the lateral regions before they had time to organize and develop their own capacity to regenerate heads, thus showing how an orderly regeneration of only one head from an anterior wound was brought about.

The inhibitory influence had to travel roughly 500 microns; if the size of a neoblast in the blastema is about 10 microns, then about fifty cells transmit the influence in, say, 80 hours, giving an average of 100 minutes per cell. Such estimated figures of the speed of transmission of inhibitory forces may be of some value if measurements of the rate of organ formation during regeneration in planarians are to be made.

Workers in planarian regeneration know that different species may behave in different ways during regeneration, giving rise to different ideas about basic problems. Caution is therefore indicated in making generalizations. Brøndsted (1954) investigated inhibition in *Bdellocephala* using a slightly altered technique. This species was selected because its head regeneration was faster than that in *Dendrocoelum*. The previous technique of making "arms" (Fig. 91) by cutting out a median strip from the decapitated animal was used to see if the connection with the main part of the body might influence the

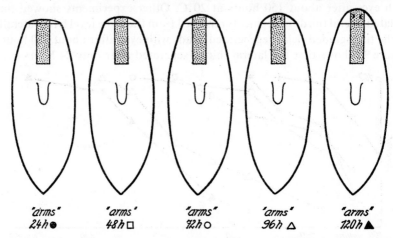

'arms'   'arms'   'arms'   'arms'   'arms'
24h●   48h□   72.h○   96h△   120h▲

FIG. 91. *Bdellocephala punctata*. Further explanation in text. (From Brøndsted, 1954.)

inhibitory mechanism. Figure 91 shows the procedure. *Dendrocoelum* required about 150 hours after decapitation to regenerate heads with eye-spots discernible at a magnification of 20 but *Bdellocephala* only took about 100 hours under the same conditions (20°). Six groups of animals were used. In the first the median strip was removed immediately after decapitation, and the "arms" regenerated heads with eyes after about 200 hours. In the second group the median strip was removed after 24 hours, and so on, to group 6, from which the median strip was removed after 120 hours. Figure 91 shows that the median "high-point" had distinct eye-spots by 96 hours and after 120 hours the regenerated heads were clearly normal. Figure 92 shows that the inhibition from the median "high-point" travels faster in *Bdellocephala* than in *Dendrocoelum*, attaining its greatest force after about 96 hours, this being in accordance with the greater regeneration speed in *Bdellocephala*. After 96 hours a curious reduction of the inhibitory force developed seen by the fact that the "arms" from group 6 regenerated heads sooner than in group 5.

The interpretation of this phenomenon suggested that when the animal had completely regenerated its head, the levels of the time-graded field were also normalized (Chapter 5), hence the "arms" after 120 hours were almost at the normal regeneration rate. The interpretation must also accept that the inhibitory forces emanating from the median "high-point" grew in strength with the growing differentiation of the blastema in the median head, but on attaining full differentiation the inhibitory forces declined to the minimum

necessary in the intact body to secure normal individualization. Figure 93
shows this interpretation schematically.

Several experiments thus show that inhibitory forces were released by
wounding and were presumably most strongly generated in the assembled

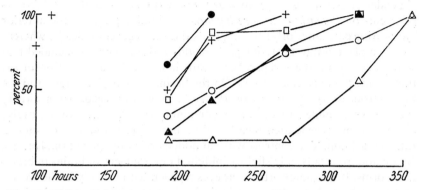

FIG. 92. *Bdellocephala punctata.* Further explanation in text. (From Brøndsted, 1954.)

neoblasts undergoing determination and differentiation. It was therefore
understandable that they travelled fastest in the blastemata made up of neoblasts and that heads may be regenerated in "windows" far distant from the
anterior blastema. If the anterior wound were connected with a much posterior one, as in the experiment described by Morgan (Fig. 86) inhibitors might
reach the posterior part in time to block head regeneration there.

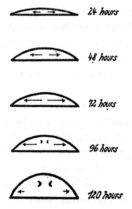

FIG. 93. *Bdellocephala punctata.* Explanation in text. (From Brøndsted, 1954.)

The necessity for postulating inhibitory forces to attain regulated morphogenesis in planarian regeneration has been stressed in various places by Brøndsted (1955). It was therefore extremely satisfying that Lender in a series of

important papers (1955, 1956, 1960) proved that inhibitory forces emanating from the brain actually exist and that they are of a chemical nature, as might have been assumed. (See also Wolff, 1962, 1963, and Wolff, Lender and Ziller-Sengel, 1964.)

Lender (1952) using *Polycelis nigra* repeated Chewtschenko's (1938) experiments on *Planaria (Dugesia) lugubris*, implanting pieces of the anterior eye-bearing part anteriorly; the implants did not regenerate heads, whereas pieces implanted into posterior parts readily did so. Lender concluded that some factor from the forepart (i.e. head) inhibited head formation in the transplant, possibly emanating from the brain. Lender (1955) removed the brains from many specimens of *Polycelis*. A brei of heads from twenty specimens was added to the water containing the brainless worms; twelve of the worms regenerated small brains, but thirteen did not develop a trace of a brain. The controls without added brei developed normal brains. Brei made from tails only slightly inhibited development of brains in brainless worms. It was noteworthy that eyes regenerated normally in the worms without a brain but cultured in brain-brei, so no toxin from the brei was responsible for the failure of regeneration in the brainless animals. The question of eye-induction was discussed in Chapter 9. Lender (1956) repeated the experiments with *Dugesia lugubris* with the same result, using the supernatant from the centrifuged brei and found that the inhibitor was present in this fluid. An important finding was that the inhibition was organ-specific. Lender (1956b) summarized his results as shown in Fig. 94. Although he found the brain inhibitors were organ-specific, he (1960) stated that they were not species-specific, because inhibitory substances from *Dugesia lugubris* also inhibited brain regeneration in *Polycelis nigra*.

Lender rightly suggested that when brei from the tail only slightly inhibited brain regeneration one might assume that an antero–posterior diffusion gradient of organ-specific brain inhibitors existed and that such a gradient would correspond with the well-known morphogenetic gradient proposed by Child.

Comparing Lender's results and suggestions with those of Brøndsted, at first glance there is some discrepancy, namely regeneration of a head in a "window" occurred both in the decapitated and in the non-decapitated animals, perhaps because the concentration of brain-inhibitors at the level of the "window" was so slight that the neoblasts assembling in the "window" were able to overcome this small inhibition and so had time to organize themselves into a new head, consistent with the existence of a time-graded regeneration field. Lender's experiments provide the clue for understanding Brøndsted's (1954) experiments: when the median brain was being formed by the neoblasts, it is reasonable to suppose that it generated inhibitory substances at the same time, the amount increasing as indicated by the diminishing rate of head regeneration of the lateral parts of the transverse segments.

Lender's idea of auto-inhibitors was further elaborated by Wolff, Lender and Ziller-Sengel (1964). Using *Dugesia lugubris* the authors decapitated the worms and implanted a brain into the anterior region (Fig. 95). In half of the cases the transplanted brain developed into the brain of the regenerated

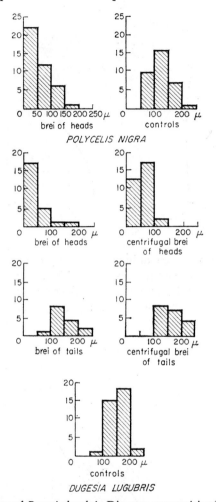

FIG. 94. *Polycelis nigra* and *Dugesia lugubris*. Diagram summarizing Lender's experiments of brain inhibitory substances. Ordinates number of animals. Abscissae size in μ of regenerated brains. (From Lender, 1956b.)

heads of the recipients; in the other half two heads formed, one regenerated from the host and one from the transplanted brain, but the head regenerated from the hosts had no brain. This experiment confirmed those of Lender, both with *Dugesia lugubris* and *Polycelis nigra*. The authors gave the outline of some further results reached by Ziller-Sengel (1965) using *Dugesia lugubris*, *D. tigrina* and *Polycelis nigra*. The results may be stated as follows: segments were isolated, made into a brei and centrifuged. Foreparts and hind parts

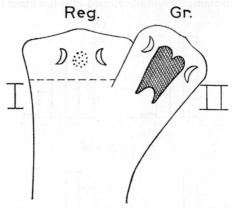

FIG. 95. *Dugesia lugubris*. The implanted brain (II) has organized a head, whereas no brain is regenerated in the blastema of the decapitated host (I). (From Wolff, Lender and Ziller-Sengel, 1964.)

were treated alike. In other animals the pharynx and surrounding tissues were excised and the wound rapidly closed. The supernatant was filtered off for use. The experimental animals were placed in this for 12 to 20 days. Figure 96 shows that the regeneration of a pharynx was considerably retarded

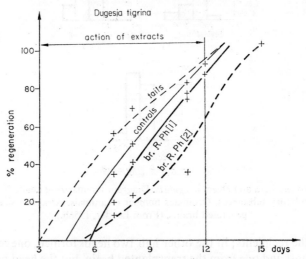

FIG. 96. *Dugesia tigrina*. Results of regeneration of pharynges in the presence of filtered extracts from tails, total pharyngeal (br.R.Ph 1) and from pharyngeal region without pharynx (br.R.Ph 2). (From Wolff, Lender and Ziller-Sengel, 1964.)

in the supernatant extracted from pharyngeal segments. It was noteworthy that anterior and posterior regeneration in the supernatant from the anterior and posterior parts respectively of the animals was generally also somewhat retarded suggesting that specific organs specifically inhibit the regeneration of their like.

The authors elaborated Lender's idea that this had some fundamental meaning, expressed in their own wording: «Une nouvelle interpretation de la théorie des gradients physiologiques dans la régénération des Planaires.» Figure 94 illustrates the idea of declining amounts of brain-inhibitors along the antero–posterior axis of *Polycelis*. The authors conclude: «On peut considérer que les gradients d'inhibition établis par CHILD sont due en grande partie à la diffusion de substances inhibitrices qui sont plus ou moins diluées suivant la distance de la source.» And later: «Nous proposons l'hypothèse de la formation et de la diffusion de substances inhibitrices comme une explication générale plausible des phénomènes de dominance et des gradients physiologiques au cour de la régénération des planaires.»

These thoughts are discussed in Chapter 23, when the whole system of organic relations in the body under regeneration will be enlarged upon. It is perhaps pertinent to refer to the papers of Saxen *et al.* (1965 and 1965) where cell contacts were stressed as part of the differentiation processes. In these papers the modern literature may be found.

CHAPTER 11

DISTRIBUTION OF NEOBLASTS
IN THE PLANARIAN BODY

ALTHOUGH the basic question as to which cell-types take part in the regeneration processes is not satisfactorily solved, it may be assumed that neoblasts play a major role (Chapter 7). It seems, therefore, very desirable to find out how these cells were distributed in the intact body ready for mobilization. In any case, a knowledge of the distribution of these important regenerative cells must be obtained for proper understanding of the regeneration mechanisms; especially as such a knowledge could be correlated with the time-graded regeneration field.

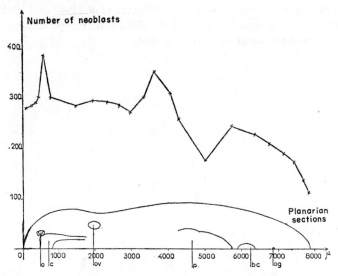

FIG. 97. *Dugesia lugubris*. Distribution of neoblasts in the resting body before regeneration. It is to be noted that the pharyngeal region is poor in neoblasts. (From Lender and Gabriel, 1960.)

Many clues as to localization of neoblasts occur in the literature, especially their topographical connection with the lateral nerve trunks, but few workers have had patience to count them systematically.

114

Lillie (1901) pointed out that *Dendrocoelum lacteum* was unable to regenerate a head from the posterior part of the body; he ascribed this to the diminishing influence from the brain. Curtis and Schulze (1924) put forward the

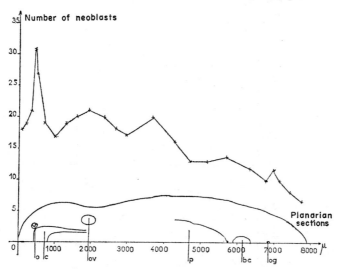

FIG. 98. *Dugesia lugubris*. Distribution of neoblasts in the resting body before regeneration. In contrast to Fig. 97 regard is taken to the fact that the pharynx do not contain neoblasts. (From Lender and Gabriel, 1960.)

hypothesis that a certain number of neoblasts were decisive for regenerative ability citing some figures for the number of neoblasts in *Dugesia tigrina, Phagocata gracilis* and *Procotylon fluviatilis* (a species closely allied to *Dendrocoelum lacteum*), viz. about seventy neoblasts in the first species during division and five at rest; twenty-two in the second species and only fifteen in the third. These numbers were given per unity, but it is impossible to see how they were obtained. Isely (1925) and Curtis and Schulze (1934) repeated the observations without giving information as to counting methods. They gave the ratio of neoblasts in *Dugesia tigrina* and *Procotylon fluviatilis* as 8·5 to 1.

It was only much later that counts were made more exactly by Lender and Gabriel (1960a, b, 1961), Stéphan-Dubois (1961) and A. and H. V. Brøndsted (1961). Lender and Gabriel used *Dugesia lugubris*, the well-known hardy species with great regenerative ability (see p. 37, Fig. 27). To discern between neoblasts and cell types the authors stained the sections with methylene-green pyronine. Figures 97 and 98 show the curve obtained by counting neoblasts in several of the sections. It is not clear, however, whether all the sections were used for counting or if the crosses on the figures indicated the sections counted; at all events, the curves cannot be far from the truth. In Fig. 99 all neoblasts in the counted sections are registered, in Fig. 100 the neoblasts in section surfaces of 114 μ^2; by this procedure the large area occupied by the pharynx, poor in neoblasts, was eliminated. It can be seen that the number of neoblasts decreased along the antero-posterior axis in accordance

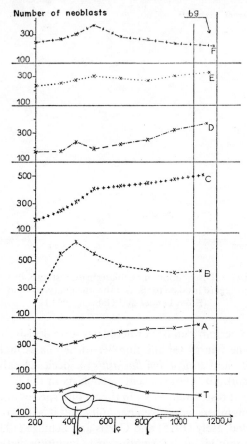

FIG. 99. *Dugesia lugubris.* Diagram of distribution of neoblasts during regeneration of a posterior region. T, control; A, 24 hours; B, 48 hours; C, 3 days; D, 4 days; E, 5 days; F, 6 days; bg, blastema; c, brain; o, eye. (From Lender and Gabriel, 1960.)

with the time-graded head regeneration field. The neoblasts were more numerous on the ventral side of the animal somewhat clustered around the ventral nerve cords, a feature also noted by several other authors. I wish to emphasize this feature, because it may indicate a close physiological and biochemical connection between nerve and neoblasts.

The authors transected the animals at the level of the ovaries; counts showed that the neoblasts migrated to the wound from the neighbourhood to form the blastema, and that neoblasts far from the wound multiplied by mitosis, and then migrated towards the wound thus enlarging the blastema. After 6 days the distribution of neoblasts was again normal.

Stéphan-Dubois (1961) undertook a similar investigation in *Dendrocoelum lacteum*, also using methylene-green and pyronine. Figure 101 shows

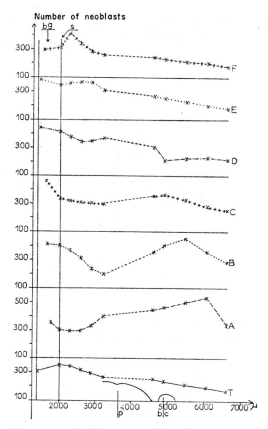

Fig. 100. *Dugesia lugubris*. Diagram of distribution of neoblasts during regeneration of an anterior region. Same abbreviations as in Fig. 99. (From Lender and Gabriel, 1960.)

the level at which counts of neoblasts (7 or 8 section of 5μ) were made. Figure 102 shows the density curve of the distribution of neoblasts per $5000\mu^2$ of surface (solid line) and the same after correction by eliminating the area occupied by the brain, pharynx and copulatory organs (hatched line). It is seen that the distribution in the forepart only fell off a little in accordance with the time-graded head regeneration field anterior to the pharynx in this species. Stéphan-Dubois did not find a concentration of neoblasts around the nerve-cords, a noteworthy observation because it might mean that neoblasts, under certain physiological conditions, may migrate in the intact body. I think that this point would repay experimental investigation concerning the physiology of these animals.

The main result of Stéphan-Dubois' paper was summarized thus:

> Le dénombrement des cellules de régénération (néoblastes) chez la planaire *Dendrocoelum lacteum* au repos montre que cette espèce, si elle possède moin de néoblastes qu'une planaire à régénération totale, est cependant bien pourvue à tous les niveaux. La perte du pouvoir de régénération céphalique dans les territoires postérieurs ne peut s'expliquer par une carence en néoblastes.

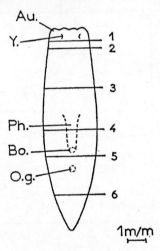

FIG. 101. *Dendrocoelum lacteum*. Au, auricles; Y, eye; Ph, pharynx; Bo, mouth; O.g, genital opening. See text. (From Stéphan-Dubois, 1961.)

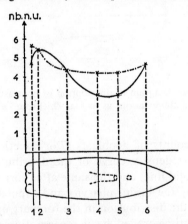

FIG. 102. *Dendrocoelum lacteum*. Number of neoblasts. Full line actual numbers, stippled line corrected for the brain, the pharynx, the copulatory organs. See text. (From Stéphan-Dubois, 1961.)

Stéphan-Dubois and Gilgenkrantz (1961a,b) found a ratio of seven neo-blasts in *Dendrocoelum lacteum* to ten in *Dugesia lugubris*, but they rightly declared that this difference did not explain the steep fall of the regenerating ability just anterior to the pharynx in *Dendrocoelum*. The authors proved irrefutably that neoblasts from all parts of the body, including the hind part, were able to regenerate a head, the latter when they had migrated through the irradiated anterior part of the body. By transplantations they proved the same, and concluded that: «... chaque territoire conserve donc ses potentia-lités propres». This confirms the experiments of Brøndsted (p. 37). The ex-periments of Stéphan-Dubois and Gilgenkrantz seemed to me to pose very important questions: how may an irradiated, and therefore doomed, part of the body still retain the power to influence traversing neoblasts to regener-ate a head? Or is it that the neoblasts themselves can regenerate a head by self-differentiation?

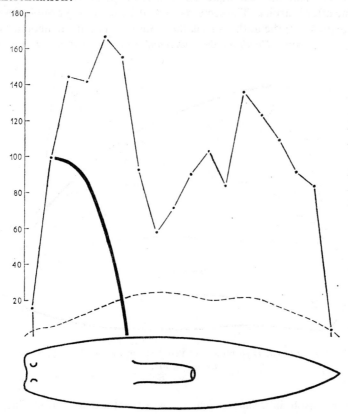

FIG. 103. *Dendrocoelum lacteum*. Thick line: head regeneration frequency curve. Stippled line: cross-sectional areas of the animal. Thin line: curve of neoblasts found. The numbers per section (10 µ) is indicated on the left scale. (From A. and H.V. Brøndsted, 1961.)

Teshigori (1963) obtained the same results as Stéphan-Dubois and Gilgen-krantz, experimenting with *Bdellocephala brunnea*. It should be recalled that this species is not able to regenerate a head from levels posterior to the base of the pharynx. A prepharyngeal piece from an irradiated animal (8000 r) was transplanted to a posterior part. Some of the animals regenerated heads from the anterior cut surface of the irradiated piece, which, if not transplanted would have died. It is for the time being unsolved how the irradiated piece, when transplanted, could live on. The chimeras were examined histologically ascertaining that neoblasts from the posterior part had migrated to the irradiated anterior one, where eye spots were seen, only after some delay; it took 35–52 days for the heads to regenerate.

It was a happy coincidence that A. and H. V. Brøndsted (1961) published a paper with the same main results as Lender and Gabriel and Stéphan-Dubois. The staining technique was nearly identical with methylene-green pyronine after Kurnick. The sections were 10 μ, obviating double countings as far as possible; the authors took the trouble to count the neoblasts in all the sections, using *Dendrocoelum lacteum* and *Dugesia torva*. Figure 103

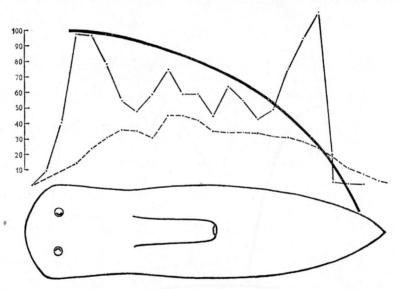

FIG. 104. *Dugesia torva*. Distribution of neoblasts in the intact body. Same indication as in Fig. 103.

shows their results in *Dendrocoelum*. It is seen that the distribution of neoblasts did not correspond with the head-regenerating ability in full agreement with Stéphan-Dubois on this major point. The specimen used was 15 mm in the living outstretched state, and contained 37,820 neoblasts, ±5%.

A specimen of *Dugesia torva* was treated in the same way. As Fig. 104 shows, this species had a head regeneration ability comparable with that of *Dugesia lugubris*. The specimen was 15 mm in the living state and 31,320 neoblasts ± 5% were counted. The authors could therefore not say that the number of neoblasts in *Dendrocoelum lacteum* was less than that of a well-regenerating species such as *D. torva*. They therefore had to discard the hypothesis of Curtis and Schulze, especially because *Dendrocoelum lacteum* regenerated both head and tail readily. Therefore the rate of regeneration was not dependent on the distribution of neoblasts in the normal body, but the size of the blastema was. In this connection it may be useful to cite the authors: "It may be of some interest to note that the total volume of neoblasts in *Dugesia torva* is about 0·0063 cu.mm., the average diameter of a neoblast (in the fixed state) being about 6 μ (the cytoplasm being rather scanty). The total volume of the inspected volume (in the fixed state) is 6·2 cu.mm, thus exceeding the volume of the neoblasts by a factor of 1·000."

CHAPTER 12

MIGRATION AND MULTIPLICATION
OF NEOBLASTS

ONE of the unsolved enigmas of planarian regeneration is the question as to which mechanisms are involved when the stimulus from the wound induces some of the neoblasts to migrate and some to start mitosis. We may be sure that these mechanisms are of humoral character, but even if we give them the insignificant name "wound-hormones", we can for the time being only hypothesize about their origin and biochemical characteristics. We do not even know if such substances originate from the nervous system, the damaged cells or from both. This is a great pity because the phenomenon is of the broadest biological significance, as it is found everywhere in all multicellular organisms, and also, in all probability, hidden in the repair processes of unicellular organisms. I think that one should keep a sharp eye upon the lysosomes as a possible source.

We can only state that in planarians neoblasts are influenced by the wound formation. It is not known at present how the wound stimulates the behaviour of the neoblasts in two quite distinct ways, that is, why the stimulus induces neoblasts in the neighbourhood of the wound to migrate with infrequent mitoses, but at the same time induces mitoses in neoblasts lying far away, which are only induced to migrate slightly? Before we try to hypothesize on these problems, let us have a look at the evidence.

1. *Migration of the Neoblasts*

It was an old conception that regenerative cells migrated to the wound or other places in the body where organ formation occurred, e.g. Wagner (1890), Lehnert (1891), Keller (1894), Morgan (1900a), Stevens (1901–2, 1907), Stevens and Boran (1905), Thatcher (1902), Stoppenbrink (1905a). Steinmann (1925) also spoke of "Wanderzellen", although he thought them formed partly by transformation of other cell types. Weigand (1930) and especially Bartsch (1923a, b) described spindle-shaped "Bildungszellen", corresponding to neoblasts, and ascribed a major role to these as wandering cells in regeneration. Curtis (1902, 1928, 1936), Curtis and Hickman (1926) and Curtis and Schulze (1924, 1934) strongly emphasized the significance of migratory neoblasts from "a persistent embryonic stock" in regenerative processes, the latter authors rather exaggerating this idea, as seen in Chapter 11. In the years following more and more authors subscribed to the view that certain cells of embryonic character migrated to the wound until the problem was finally settled by Dubois (1949), as pointed out in Chapter 7. Since then an imposing array of supporting evidence has been given, especially by the Wolffian school. Conclusive proof was given: Kolmayer and Stéphan-Dubois (1960) and in two papers by Stéphan-Dubois and Gilgenkrantz (1961a, b) which in addition gave additional important information.

The authors used *Dendrocoelum lacteum*, a species unable to regenerate a head from levels behind the radix of the pharynx. The authors began with an experiment similar to that of A. and H. V. Brøndsted (see Chapter 11) using *Bdellocephala punctata*, a species showing the same lack of head-forming ability behind the pharynx (Fig. 20). Figure 105 shows their first experiment in which the animals were decapitated; *A* shows the prepharyngeal piece

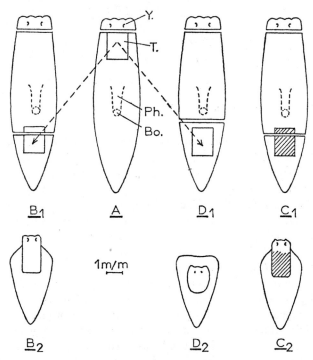

FIG. 105. *Dendrocoelum lacteum*. Explanation in text. (From Stéphan-Dubois and Gilgenkrantz, 1961b.)

excised; in *B* this piece was transplanted instead of an excised posterior piece, a transverse cut being made through the graft, which alone regenerated a head, B_2. The same experiment was performed as shown in D_1, but differing in that the transverse cut was made a little anterior to the transplanted graft;

D_2 shows that the graft, and this alone, regardless of the procedure, regenerates a head. The crucial experiment is in C_1. A prepharyngeal piece from an irradiated specimen was grafted to a non-irradiated animal as in B_1; but a transverse cut was then made through this graft, the neoblasts of which had been killed by irradiation (Dubois, 1949, 1951); the graft was therefore unable to regenerate anything on its own, though C_2 shows that it nevertheless regenerated a head. The only possible explanation is that neoblasts

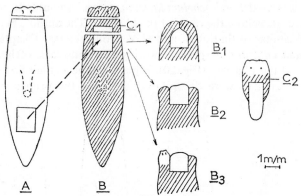

FIG. 106. *Dendrocoelum lacteum*. Explanation in text. (From Stéphan-Dubois and Gilgenkrantz, 1961b.)

from the host's tail region migrated through the irradiated graft and regenerated a head. Therefore the graft had retained its anterior potentialities in the regenerative field of this species; furthermore, it had retained its inducing powers, perhaps because the nervous system had not yet been damaged by the irradiation. The reciprocal experiment was performed as shown in Fig. 106. In A, a postpharyngeal piece was transplanted to an irradiated animal which was therefore devoid of living neoblasts. After decapitation and being cut transversely through the graft, three situations may occur, B_1, B_2 and B_3. In the latter a new head was regenerated, but laterally which may be interpreted by assuming that neoblasts from the non-irradiated graft migrated to the front of the irradiated host and were there induced to regenerate a head. C_2 shows that the anterior piece cut out in C_1 and containing the forepart of the graft was then able to regenerate a head, again because it contained intact neoblasts from the graft (Fig. 107).

These experiments were important because they demonstrated two main points: (1) that neoblasts migrated and (2) in the author's own phrasing: «... chaque territoire conserve donc ses potentialités propres». This was of great significance, because it confirmed the idea of the existence of a time-graded regenerative field (Chapter 5). The neoblasts including those in the tail region were totipotent; this disproved conclusively Goetsch's idea that two kinds of regenerative cells existed, regenerating head and tail respectively.

Stéphan-Dubois and Gilgenkrantz (1961b) repeated these experiments (Fig. 107) and extended them by giving the time taken for head regeneration in the graft. We know (Chapter 11) that the number of neoblasts in the tail part of *Dendrocoelum lacteum* is less than in the anterior region, therefore the regeneration of a head from an irradiated anterior graft to a tail takes 2 weeks instead of the normal 9 days.

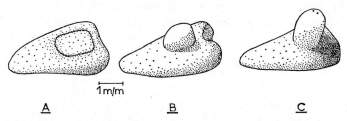

1 m/m

A   B   C

FIG. 107. *Dendrocoelum lacteum*. Explanation in text. (From Stéphan-Dubois and Gilgenkrantz, 1961a.)

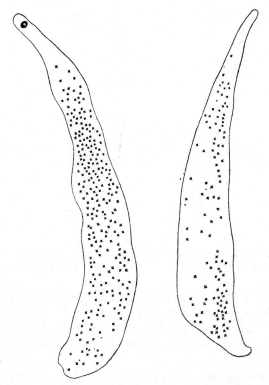

FIG. 108. *Polycelis nigra*. Distribution of mitoses in anterior half (left) and posterior half (right), 1 hour after operation. (From Verhoef, 1946.)

2. *Multiplication of Neoblasts*

Although Dresden (1940) could not find mitoses in regenerating planarians, most authors agree that mitoses are found. Statements have been given in several places in this book but few exact counts have been made.

Verhoef (1940), using the Feulgen technique, found an equal distribution of mitoses throughout the intact animal and also up to 2 days after transection (Fig. 108). No diurnal periodicity was detected. Animals transected through the pharyngeal part showed (Fig. 109) the number of mitoses increasing near the wound 84 hours after the operation.

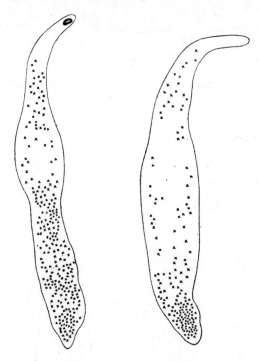

FIG. 109. *Polycelis nigra.* Distribution of mitoses in anterior half (left) and posterior half (right) 84 hours after operation. (From Verhoef, 1946.)

Pedersen (1958), in his paper concerning the influence of TEM (p. 65) on the regeneration of *Phagocata vitta*, found that in the controls (animals regenerating under normal conditions) the number of mitoses rose very considerably, from 0·79 per 1000 to 4·71 per 1000 on the third to fourth day after decapitation.

McWhinnie and Gleason (1957), in *Dugesia dorotocephala*, found that the parenchymal amoebocytes (neoblasts) were the only ones showing mitoses during regeneration, the peak being on the third day. They found, as did

Dubois (1949), that the neoblasts were the only cells which replaced lost parts and that colchicine inhibited their mitotic activity.

Lindh (1957c) dealt meticulously with the mitotic activity in *Dugesia polychroa* (cloned animals). Like Verhoef he found an even distribution of mitoses in the intact animal. Lindh wrote:

> ... but as soon as a part of an animal is forced to regenerate, a graded distribution appears both within the different longitudinal parts and along the old axis, if the number of cell divisions in the part is compared. The greater mitotic activity occurs just behind the old mid-part, from which point the activity decreases cephalically and caudally. However, the activity in the old head is next to the mid-part, maximal.

Lindh found curious rhythms in the mitotic activity, described in his paper. He also found many significantly different variations in the mitotic spectrum between young and old animals (Fig. 110). Figure 111 shows the distribution of mitoses in the posterior regeneration from a head cut away from the body.

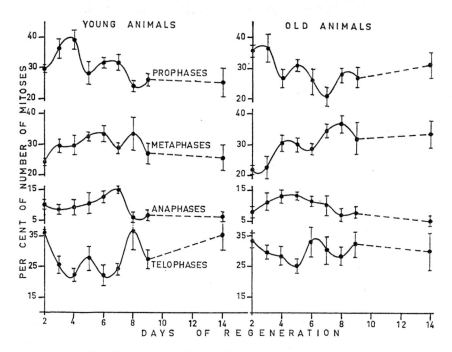

FIG. 110. *Dugesia polychroa.* See text. (From Lindh, 1957c.)

Regarding the cited papers there can be no doubt whatsoever as to the increase in mitoses on wounding the animal, and therefore of the significance of mitoses in regeneration. But we are still ignorant of the biochemical mechanisms (wound-hormones?) which elicit mitoses on wounding. Perhaps a careful analysis of the biochemical events occurring with the use of various chemicals to suppress mitosis (e.g. Pedersen, 1958, with TEM, and Kanatani, 1962, with nicotinamide) might in the future enable us to understand what "wound-hormones" really represent.

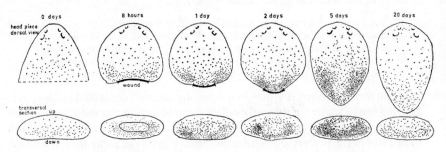

FIG. 111. *Dugesia polychroa.* See text. (From Lindh, 1957c.)

CHAPTER 13

THE NERVOUS SYSTEM
AND REGENERATION

In Chapter 9 we dealt with the ingenious experiments of the Wolffian school, especially those of Lender. We saw that the head ganglia were necessary for eye-formation. But we may still ask: Is the nervous system in fact the decisive promotor for overall regeneration? In other words: Is planarian regeneration impossible without the presence of the nervous system? One thing, however, we must be keenly aware of: if the nervous system is necessary for starting regeneration, and if of the nervous system, the brain at least, is necessary for the regeneration of eyes and perhaps other organs, the reverse is certainly not true because if the nervous system is intact full regeneration will not necessarily follow automatically in all cases. I refer again to my favourite problem: why does *Bdellocephala* and *Dendrocoelum* fail to regenerate heads from anterior wounds behind the pharynx, in spite of plenty of neoblasts and the presence of the lateral nerve trunks and other nerves?

The literature concerning the necessity of the nervous system for regeneration is somewhat conflicting, although most authors have held that it is a prerequisite for regeneration. Let us have a brief look at the older literature.

I have already quoted Dugès (1828), who gave good reasons but no proof for its necessity. Bardeen (1902) recorded the absence of well-defined nerve cords in the hind part of very young animals ("embryos") of *Dugesia dorotocephala*, and thought this to be the reason why they did not regenerate heads from the hind part; older animals regenerated heads readily from the same levels. Bardeen deduced from this that the nervous system was necessary for regeneration and also decisive for polarity, though his reasoning was not conclusive. In the first place, what did he mean by the expression "absence in the piece of well-formed nerve cords"? One would think that Bardeen had found nerves. Secondly, it was conceivable that the hind part in these young animals was still unable to be induced to regenerate heads. The problem is well worth investigating.

T. H. Morgan (1905) found, in the polyclad *Leptoplana*, that cephalic ganglia were necessary for good regeneration; on excision no regeneration occurred on the surface of pieces behind the cephalic ganglia. But L. V. Morgan (1905) and Olmsted (1922), also using *Leptoplana*, maintained that regeneration occurred without head ganglia but was quantitatively poorer. They held that the influence of ganglia for good regeneration was only indirect: good regeneration demanded locomotion, and locomotion required head ganglia. This too is also a problem well worth reinvestigation.

Levetzow (1939), using various polyclads, maintained that regeneration occurred qualitatively without head ganglia, but that regeneration was quantitatively poorer. Schultz (1902b), also working with polyclads, did not find regeneration posterior to the head ganglia, but he simply ascribed this to closure of the wound. I should think it worth while to repeat these experiments by reopening the closure by two oblique cuts meeting in the midline to prevent further closure. This procedure may be used with advantage in many cases (Fig. 112). Child (1904d, 1910b) was in agreement with Levetzow.

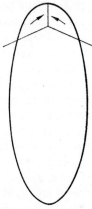

FIG. 112. Diagram of cutting to avoid closure of wound. (Original.)

Santos (1929, 1931), discussed in Chapter 9, held that sections of the ganglionic region if sufficiently large may induce post-cephalic outgrowths, head formations and pharyngia when grafted prepharyngeally. Wilson (1941) held that the blastema only differentiated because nerves from the body induced it. Silber and Hamburger (1939) were of the opinion that stimuli from branches of the nervous system, when exposed by lateral wounds, induced head formation. Beyer and Child (1933) did not regard the nervous system as an essential factor in regeneration, although they thought that it might have some influence on the physiological conditions of the surrounding tissues. This somewhat cryptic notion may conceal an important truth, but only investigations with modern techniques can reveal it. The problem is, however, so essential that it certainly ought to inspire such investigations. Child (1941) later emphasized the important role of the nervous system in regeneration.

It is a remarkable feature of Child's metabolic gradient hypothesis that the nervous system is sometimes regarded as a sort of inhibitor of regenerative processes, at least Child and Watanabe (1935) and Watanabe (1935b) regarded the head-frequency somewhat depressed by the nervous system, because they found an increase in head-frequency by blocking the nerve cords by potassium cyanide and anaesthetics.

In *Rhabdocoels* Fulinsky (1922) found repair but no proper regeneration, the brain not playing any role in the reparative processes.

The controversial opinions in the older literature are not much help and it is therefore satisfying that some recent papers seem to indicate possible ways for clarifying the problem. Bondi (1959), using *Dugesia lugubris*, cut the animals as shown in Fig. 113 and obtained a regenerate with a very shortened

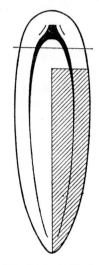

FIG. 113. *Dugesia lugubris*. See text. (Redrawn after Bondi, 1959.)

lateral nerve cord on the one side. It was essential that the anterior blastema formed in the decapitated worm was always symmetrical, but the eyes were regenerated at very different speeds, the one corresponding to the longer nerve cord appearing 3 days before that corresponding to the shorter. Bondi held that the cerebral ganglia were formed by an epimorphic proliferation of the ventral cords and not from neoblasts, in striking contrast to the opinion of the French school. It was at all events significant that the larger nerve cord seemed to promote eye formation at a higher rate than a short one; but it is also significant that the blastema was symmetrical. It is therefore reasonable to suppose that the migration of neoblasts was independent of the nervous system, and induced eye formation, as Lender has shown.

Lender and Gripon (1962) confirmed Bondi's finding, but found that the retardation of eye formation due to the shorter nerve cord was only 24 hours. They found that the brain was somewhat smaller here, but they thought that head ganglia were regenerated by neoblasts («à partir des néoblastes») as prolongation of the nerve cords.

We know from Lender that induction of eye-formation is of a neuro-humoral character, but these experiments do not enlighten us as to the general starting-point for regeneration. Stéphan-Dubois and Lender (1956) have shown that only a real wounding started the migration of neoblasts, whereas cauterization, electrocoagulation, irradiation and other forms of trauma did not evoke such a migration. The authors therefore concluded that wound-hormones triggered off the migration of neoblasts, and suggested that these hormones may be neurohormones. If so, the nervous system could have two functions in regeneration: an overall one (initiating migration of neoblasts) and a special one (induction of specific organ formation, at least of eyes). It is noteworthy that these authors found eyes to regenerate before the nerve fibres from the brain reached them; the induction was therefore thought to be humoral, by "organisin" secreted from the brain.

It is now likely that Lender and Klein (1961) by histochemical staining (Gabe's fuchsin paraldehyde and Masson's hemalum and picro-indigo-carmine) have found certain ventral secretory cells in the brain (Figs. 114,

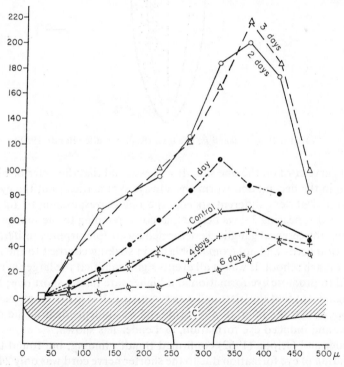

FIG. 114. *Polycelis nigra*. Distribution of secretory cells in the brain during regeneration of the posterior part of the worm. Abscissa: length of brain. Ordinate: number of secretory cell per 50 μ of the brain. C, brain with the connecting commissure of the two brain halves. (From Lender and Klein, 1961.)

115). When the animals were cut prepharyngeally these cells underwent alterations in shape, and increased the size of their nucleoli; moreover, their numbers increased because ordinary brain cells transformed themselves into secretory cells, and the stainability of all secretory cells increased. These important findings were enlarged upon by Lender (1964). He found small cells

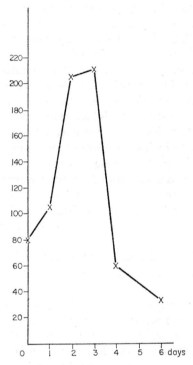

FIG. 115. *Polycelis nigra.* Variation of number of the secretory cells in the brain during 6 days of regeneration of a posterior part of the worm. Ordinate: average number of these cells plotted to 100 μ of the brain. (From Lender and Klein, 1961.)

in *Polycelis nigra, Dendrocoelum lacteum, Dugesia lugubris, D. tigrina* and *D. gonocephala* lying dorsally and posteriorly to the brain cells (Fig. 116) about 11–20 μ in size, with small nuclei, and secretory granules located along the axons going to the parenchyma and epidermis. During regeneration the number of these cells increased, attaining a maximum on the third day when the brain appeared in the posterior part of the body regenerating the head though no neurosecretion was demonstrated. The author concluded that inductive and inhibitory substances: «... ne sont donc pas assimilables à la neurosecretion...». But neurosecretions appeared at the same time as the copulatory organs developed. Lender found that in *Polycelis nigra,* where cocoons are found during the whole year, neurosecretions were always present. In *Dendrocoelum* cocoons are found in September, in *D. lugubris* and *D. gonocephala* only in spring, and only then were neurosecretions found. The neurosecretion disappeared when the copulatory organs involuted. Lender suggested that the neurosecretion «... favorise le régénération d'un organe de la région caudale, par example, l'àppareil génital?»

These findings indicated broad perspectives: what kind of secretion is made by these cells? Is the secretion able to initiate the migration of neoblasts? Is Lender's "organisin", the inductive substance for eye-formation, in reality the secretion from the secretory brain cells?

Török (1958) indirectly attributed an overall influence of the nervous system on regeneration by suggesting that the number of nerve cells was connected with rate of regeneration in the time-graded head regeneration field (Chapter 5). It is hard to see how this idea may be reconciled with the ability to regenerate tail but not head from the posterior part of *Bdellocephala* and *Dendrocoelum*, perhaps indicating another mechanism inhibiting head formation in the posterior parts of these genera.

I think that we have to understand the biochemistry of the nervous system thoroughly before we are able to say anything definite about its influence on the initiation of regeneration and remoulding of the tissues by neoblasts and perhaps other tissue cells. A beginning has been made by some research writers.

Yamamotu (1957) found alkaline phosphatase in the brain (Fig. 117). Gazsó, Török and Rappay (1961), using *Dugesia lugubris* and *Dendrocoelum lacteum*, found alkaline phosphatase in the whole nervous system, localized in the nuclear membrane, nucleoli, neurofibrils and neuromuscular connections. The specific cholinesterase was only found in somatic neuromuscular junctions of the motor nerve ends. It was remarkable that there were no positive reactions in *Dendrocoelum lacteum*; the authors suggested that this might be due to the lower regenerative ability of this species. This is doubtful, especially in view of the very good capacity for regeneration in the tail of this animal.

Ichikawa and Ishii (1961) observed that the nerve cords during regeneration in *Dendrocoelopsis lacteus* grew thicker and more loose, and, assisted by neoblasts, regenerated new nerve cells. They also found alkaline phosphatase in the intact animals in the brain and several other organs. The activity of alkaline phosphatase increased in the nerve cords during regeneration. The authors thought that the phosphatase was probably generated in the brain.

It would be premature to draw conclusions from these sparse biochemical data in relation to our main problem, nervous system regeneration, but we have to regard them as stones in the future mosaic picture of planarian regeneration.

Kido's (1952) remarkable experiment may on expansion help to clarify both the problem of dorso-ventral polarity and the influence of the nervous

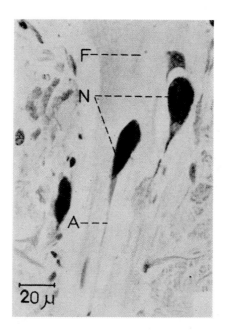

FIG. 116. N, secretory cells; A, axons; F, nervous fibres. (From Lender, 1964.)

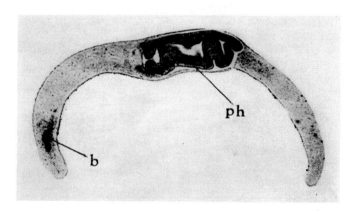

FIG. 117. *Dendrocoelopsis* sp. Section showing the distribution of alkaline phosphatase. b, brain; ph, pharynx. (From Yamamotu, 1957.)

system. Using an asexual strain of *Dugesia gonocephala* he removed a section from the post-cephalic region (Fig. 118) and divided it into a dorsal and ventral part, the latter including parts of the ventral nerve cords. The pieces were grafted on to the postpharyngeal dorsal side of the host. Only the ventral pieces containing the nerve cords induced new tissues and were not resorbed, like many of the dorsal pieces. Moreover, the ventral pieces induced pharynx formation in all cases, whereas only two out of twenty-two dorsal pieces did so.

It is, of course, interesting to postulate a possible connection between the nervous system and regeneration in other animal groups, but we must not generalize too freely, and only a few examples will be cited.

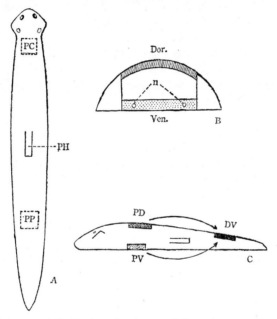

FIG. 118. *Dugesia gonocephala.* Explanation in text. PC, graft removed to PP; n, nerve cords; B, cross-section of post-cephalic region, dorsal graft hatched. Ventral graft dotted. (From Kido, 1952.)

Penzlin (1954) working with limb regeneration in the larval stages of *Periplaneta americana* says: "Es gibt keine Hinweise dafür, daß eine Wiederherstellung der Nervenverbindung (after denervation) Voraussetzung für die Regeneration ist." But he also states: "Das Wachstum des Beinregenerates nach Denervation is gegenüber der Kontrolle verlangsamt," and: "... seine Muskulatur is spärlich entwickelt, es wird vom Tier nicht bewegt."

In vertebrates Singer (1947, 1960) and Singer and Mutterperl (1963) found that nerve fibres were necessary for the regeneration of limbs; in their 1963 paper the authors stated: "The wound tissues give rise to the new growth. They determine the nature and growth characteristics of the regenerate, whereas the role of the neurone is a relatively non-specific one, providing some factor important for the growth."

It seemed that this mechanism contrasted somewhat with the situation in planarians, where Lender found a specific role for the nervous system, the brain's induction of eyes. Such differences should stimulate further researches as to the hypothetical overall significance of the nervous system in planarian regeneration.

In view of the increasing body of evidence that RNA has a major role in the various activities of the nervous system, the findings of A. and H.V. Brøndsted (1953), that RNA (from yeast) accelerated regeneration in starved planarians, are perhaps relevant. It is probable that the yeast RNA is degraded to nucleotides by cells near the wound, and that the starving nerve cells build up the nucleotides to specific RNA, thus speeding up the regeneration processes with neurohormones. But it is also possible that all cell types, especially neoblasts, are invigorated by the RNA. Further investigation into the incorporation of RNA by isotopic methods should prove enlightening.

Even if it were proved that the nervous system was responsible for migration of neoblasts, inducing formation of eyes, and perhaps other organs, we are still ignorant as to the basic problem: how is the nervous system prompted to do these things; what are the basic trigger mechanisms when a wound is made? We shall revert to these questions in the concluding Chapter 23.

CHAPTER 14

HETEROMORPHOSES
AND REINDIVIDUALIZATION

ONE of the riddles of regeneration is that structures different from the normal organs are sometimes regenerated. Well-known examples are given in textbooks, e.g. the regeneration of limbs instead of antennae in arthropods. Such abnormalities were called *Heteromorphoses* by Loeb (1891). In planarians heteromorphoses may be very grotesque, so refuting vitalistic concepts as "Entelechie" (Driesch, 1908), "Planmäßigkeit" (Steinmann, 1926), or directiveness (Russell, 1945). In Chapter 6 we discussed the heteromorphoses called "Janus-heads" and "Janus-tails", emphasizing that such heteromorphoses were found in the nature without experimental participation. A very interesting instance was given by Jenkins (1963) when she found in a stock culture of *Dugesia dorotocephala* four specimens with bipolar heads; the specimens were 30 mm long. They could not be related to stimuli from either temperature or chemicals. It is noteworthy that the strain was an asexual one.

The literature dealing with experimentally induced heteromorphoses is extensive, for example, Hallez (1892, 1899, 1900), Van Duyne (1896) and Randolph (1897), who had already obtained by median splitting methods double-headed animals and animals with double pharynges, Bardeen (1903) (several fine experiments), Sekera (1911), Bömig (1913), Pzribram (1921) and Korschelt (1927). I shall now quote some experiments to provide material for speculating on the major features of planarian regeneration.

Goetsch (1921, 1922), using *Dugesia lugubris*, made a median cut from the tail tip to near the head; if coalescence were prevented a head might regenerate in the angle, but directed posteriorly. If after the median splitting a T-cut was made, a head might regenerate from the exposed anterior parts at the

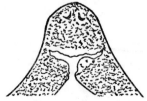

FIG. 119. Head formation following T-cut. (From Brøndsted, 1955, after Goetsch, 1921, *Arch. Entw. Mech.* **49.**)

T-cut (Fig. 119). This kind of experiment was supplemented by Beissenhirtz (1928), using *Dugesia gonocephala* and *Polycelis nigra*. Figure 120 shows the procedure and the result. The head with seemingly reversed polarity regenerated from the angle of the cut, its brain connected with the lateral nerve trunks previously regenerated in the tissue originally sited in the median parts of the animal. We may here note that if the median cut had severed the

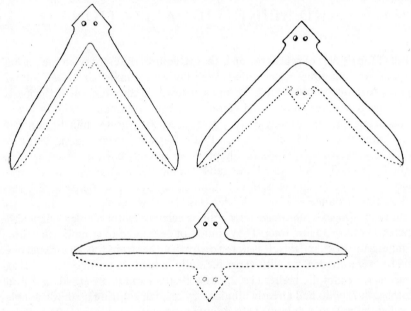

Fig. 120. Explanation in text. (From Brøndsted, 1946, after Beissenhirtz, 1928.)

two halves completely, each half would have fully regenerated the missing half. Beissenhirtz cut this heteromorphosis in three ways; if he cut as indicated in I, Fig. 121, the two half-heads coalesced and formed one head. If cut as in II, each head swung into a position continuous with the main axis of the rest of the body, so regulating a normal animal. If the cut was made as in III

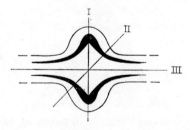

Fig. 121. See text. (Redrawn after Beissenhirtz, 1928.)

the two halves coalesced to form a normal animal in the same way as an animal split from the tail tip to near the head would have its two halves coalesce if they were not constantly prevented from doing so.

By making the same T-cuts as Goetsch, Beissenhirtz succeeded in getting two heads to regenerate from each of the exposed median angles at the T; the lateral halves regenerated their symmetrical halves, producing very curious heteromorphoses (Fig. 122), thus also disproving inhibition from the old head.

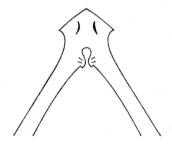

FIG. 122. See text. (Redrawn after Beissenhirtz, 1928.)

It is noteworthy that Beissenhirtz found plenty of malformation in pharynges and other organs, frequently including supernumerary eyes. He also found organiziation which was in accordance with results obtained by Steinmann (1910, 1927) indicating "reindividualization".

Goetsch (1921) succeeded in making headless twins by splitting the animals with a median cut to near the head; after regeneration of the symmetrical halves of each half-part the original head was cut off by a V-cut; the two exposed wound surfaces coalesced so that no blastema formed, and Goetsch pointed out that eyes only occurred in blastema tissues. But when he again cut off the anterior part by a transverse cut, a blastema developed forming a head with eyes.

Goldsmith (1933, 1934, 1940) used cauterization with heated platinum needles, electrical stimulation and radial incisions pre- and postpharyngeally dorsally in *Dugesia tigrina* and *Procotyla fluviatilis*. From the wound in *D. tigrina*, but never in *Procotyla*, oblong outgrowths appeared, often regenerating heads with eyes, more frequently pre- than postpharyngeally. The outgrowths often migrated anteriorly or laterally, and persisted for several months; they were often resorbed but the eyes persisted. Goldsmith was of the opinion that the stimuli initiated the migration of neoblasts and further postulated that the stimuli released—SH groups.

In contrast it may be noted that Hull (1938) attained the same heteromorphoses as Goetsch and Beissenhirtz in *Dugesia maculata (= tigrina)* and *D. agilis.* He severed the body further into three longitudinal strips by two longitudinal cuts, which did not, however, sever the head or tail. This resulted in a spontaneous outburst of heads or "cephalization", up to eight heads may be regenerated. Organization took place later.

Kahl (1938) induced heteromorphoses by sheaf cuts. His interpretations were formal and without biological foundation; but the experiments were notable because they stimulated speculation concerning the major problems of migration of neoblasts, induction, inhibition and polarity.

Keil (1924), using *Polycelis nigra*, studied heteromorphoses following splitting cuts. He rightly pointed out that the regenerates were determined by internal factors of unknown character and not by functional conditions (movements of the body).

Lus (1926) in the marine triclade *Procerodes lobata* and *Cercyria papillosa* produced several forms of heteromorphoses by cutting the animals in different ways. Postpharyngeal pieces never regenerated heads but often heteromorphic tails; short prepharyngeal pieces often regenerated heteromorphic heads. If pieces were cut out just in front of the base of the pharynx they often regenerated a tail or both tail and head from the anterior surface. Longitudinally split animals regenerated worms with two heads or two tails. Lus thought that totipotent cells were found everywhere in the body, but their determination depended upon the localization of certain formative stimuli: "Kopfreize" were absent in the hind part of the body, whereas all sorts of "Reize" were found in the forepart where the "Kopfreize" were stronger. He emphasized that these "Reize" were differently distributed in the different planarian species. It thus seems that Lus thought that two opposite gradients existed, a head-determinig gradient tapering posteriorly and a tail-determining one tapering anteriorly. This seems to conform with the situation found in *Bdellocephala* (Fig. 20, p. 32).

Rand and Ellis (1926) made a series of thought-provoking experiments on *Dugesia maculata (= tigrina)*. Heteromorphoses with either double heads or tails were formed by partial median splitting. The heteromorphoses with double tails were handled in the following ways: one tail was cut away leaving a short stump from which a new tail regenerated; after 16 days this tail was cut away close to the body; after a further 7 days no sign of regeneration was found. A V-cut was then made, and from both exposed surfaces tails regenerated; the authors concluded that the same tissues which earlier could not regenerate a tail could then do so. With double-headed animals, one head was removed close to the body and no head regenerated. If a little stump was left after cutting, still no head regenerated. If the remainder was then cut away with a cut which also transected the stump, then a new head was promptly regenerated from the wound. The authors interpreted the results thus: the inhibition of new tails or heads in double heteromorphoses "... is specifically due to the presence in appropriate proximity and relations, of the *equivalent* of the structures which have been removed... there is demonstrated a 'dominance' but an inhibitory dominance—of one part of an organism over another." (See also Chapter 10.)

Silber and Hamburger (1939), in *Dugesia tigrina*, gave curious pictures of the enormous regenerative ability of this species. Poor regeneration was due to lack of or bad development of a wound surface. The authors declared that the regenerated heads neither stimulated nor inhibited one another and probably the cut stimulated the tissue in such a way that a high-point was reached in the gradient system. The authors found difficulty in understanding the regeneration of heads more posteriorly at the lateral side of the wound on the basis of the gradient hypothesis and that perhaps the heteromorphoses were due to the influence of exposed branches of the nervous system.

Wilhelmi (1909) obtained heteromorphoses with double pharynges after extirpation of the original one and thought that similar heteromorphs found in nature might have been due to severe damage of the pharynx during feeding.

Amongst the phenomena of heteromorphoses the features of supernumerary eyes should be mentioned. Several examples are given elsewhere in the text, e.g. those of Child and co-workers, and Brøndsted (1942c, LiCl produced). Ghirardellei and Tasselli (1956) performed certain operations and found that the eyes were anomalous and supernumerary in such a way that the phenomena conformed with the existence of the time-graded regeneration field. These findings were confirmed by Kanatani (1958a, d).

Goldsmith (1941) found heteromorphoses in clones produced artificially in *Dugesia tigrina*: animals were produced with triple or quadruple heads, those with "fused" heads having up to twelve eyes, others with two to six tails and some with up to six pharynges.

Stéphan (1965) found the same in *Dugesia tigrina* with some of the animals transmitting the anomaly to the next generation; it was generally also transmitted to the fission products and to artificially separated segments from behind the pharynges.

It would be interesting to see if these anomalies were due to unusual assemblages of neoblasts tending to divide the animals longitudinally and to see how the nervous system behaved in these heteromorphs.

It has already been pointed out that coalescence often happens in heteromorphs resulting in normality. This *reindividualization* has especially been studied by Steinmann (1926, 1927, 1928), who also reviewed the older literature. Beissenhirtz (1928) presented cases of reindividualization. Steinmann used the classical method of splitting to attain double-headed heteromorphoses and found that some of these coalesced to form a single more or less normal individual; these experiments were varied in several ways.

Steinmann used this rather striking phenomenon to support his vitalistic views on morphogenesis. Everyone who followed such reindividualizations in his own experimental series must admit that one gets the impression that some weird force was at work. Steinmann concluded that "Planmäßigkeit" in the planarian tissues enforces normality and maintained that the vitalistic view was just as scientific as the mechanistic one. By clinging to the vitalistic view one isolates oneself from more penetrating research, because *eo ipso* they must be futile as one cannot get at the core of "living processes" by adhering to preconceived ideas. The mechanistic view is neither a preconceived idea nor a "Lebensanschauung", only working hypothesis, regardless of the ultimate goal. I hope that this will stimulate young colleagues to tackle the problem of reindividualization.

The problem is tantalizing; take a heteromorph, e.g. with a lateral head which after some time grows larger with a body attached to it, and let this new individual be nipped off, as seen after some months in several experiments. This may be regarded as similar to fission, a phenomenon which—let it be said at once—is also rather obscure (see Chapter 16). But the opposite may also happen: a reindividualization either by resorption (the one individual being "stronger" than the other) or by coalescence (the duplicates being of "equal strength"). This intriguing problem deserves deeper analysis.

Steinmann did not deal very much with such analyses beyond a rather careful morphological description. Later workers do not seem to have tackled the problem of reindividualization, at least, I have no knowledge of such work. This is beyond doubt due to the fact that organization during regeneration has been the major subject in the science of regeneration. I have a feeling that a penetrating analysis of reindividualization might provide some good clues for understanding regeneration itself, because such ideas as competition, dominance, inhibition and organization are involved.

Imagine a two-headed planarian with the "necks" growing shorter until the sides of the heads meet. The eyes, now situated near the mid-line and nearly touching one another, are either absorbed or persist as reduced eyes in the midline of the now reindividualized animal. Steinmann has shown that anastomoses between the intestines of the two "animals" occurred until the intestine was restored to normality.

Clearly, we cannot use the idea of "competition" here; both animals were of equal size and vigour and the two sets of symmetrical organs simply relinquished their median halves—why? To answer this it is necessary to discuss the remoulding forces involved here in a broader aspect and we must therefore leave the problem for the moment until the general discussion in the concluding Chapter 23.

It is, however, pertinent to point out a few facts regarding the individualization problem. Figure 85 shows that a head transplanted to the cephalic portion of *Bdellocephala*, but at the side of a transplanted tail, did not result in reindividualization. Ax (1957) discussed the problem of multiplication of the male copulatory organs in Turbellaria, and posed the question in relation to zooids. Stéphan (1965), in *Dugesia tigrina*, studied multiplication of pha-

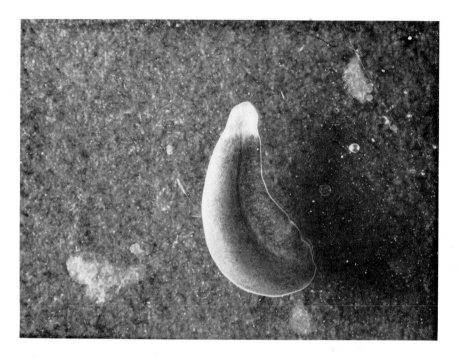

FIG. 123. Longitudinal chimeras of *Dugesia polychroa* + *D. lugubris*. See text. (Original)

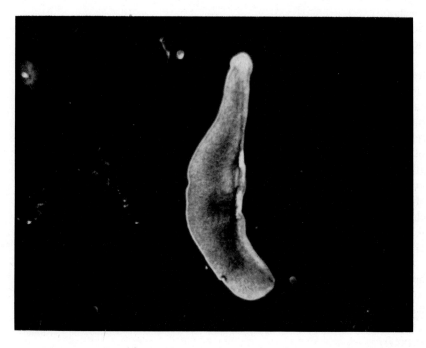

Fig. 124. Longitudinal chimera of *Dugesia polychroa* + *D. lugubris*. See text. (Original.)

rynges lying laterally to each other in strains with dorsal pigment streaks corresponding to the pharynges, and also often to lateral multiplication of eyes. He found that double pharynges were sometimes only double in the anterior parts but single posteriorly. Some animals transmitted these anomalies to the next generations arising by fission, and also to artificially separated segments from behind the pharynx. It seems to me that there must be a mechanism—perhaps due to anomalous collections of neoblasts—tending to divide the animal longitudinally. It would be very interesting to see how the nervous system behaved in such semi-double animals.

In the meantime a few words and figures will be given from as yet unpublished investigations in progress. Two decapitated longitudinal halves of the closely allied species *Dugesia lugubris* and *D. polychroa* were transplanted to each other, so building a hetero-parabiosis. After firm coalescence, the parabiosis was again decapitated and about a week later a symmetrical head regenerated from the joint blastema (Fig. 123), one half from *D. lugubris*, the other from *D. polychroa*. Later the *polychroa*, grew slowly but steadily laterally along its whole length, "pushing" the *lugubris*-half aside, indeed not only pushing but resorbing more and more of the median part of it. Eventually the *lugubris* component left was only a border on the side of the polychroa (Fig. 124). Pending closer analysis, I might propose the hypothesis that serological factors of the cells in the *polychroa*-part overcame those of *lugubris*.

Could it be that serological factors were involved in reindividualization of heteromorphs so that only one set of organs were quantitatively and qualitatively possible, the serological factors starting their work from the normal base and working outwards?

The problem of reindividualization is akin to the phenomenon of regeneration from the mucous capsules described by Alexander and Price (1926-7) and Castle (1927, 1928), both working with *Planaria velata*. In this species asexual reproduction and detachment of fragments from the posterior part leads to encystment of these fragments. Alexander and Price found a degeneration of internal organization with all organs disappearing and a mass of granular material filling the cyst with very few cells visible. The authors thought that the granular material served the cells as nutrient matter during their differentiation to form a new individual. He found the animals in a spring at Valparaiso, Indiana, and stated that good nourishment and a rise of temperature brought about encystment. At 18–20 °C encystment occurred when the animals were only 12–14 mm long, but at 8–10 °C, when 18–20 mm long, he described complete histolysis of all internal organs preceding fragmentation of the worms, leaving only a syncytium with few nuclei. Encystment into mucoid capsules ensued. Total de-differentiation led to completely disorganized material, with "... no morphological evidence of the original axes of the worm other than the difference between dorsal and ventral surfaces .... Castle thought that the original axes probably persisted throughout the process of fragmentation and encystment. The gut and the primordia of the cerebral ganglia were the first to appear.

The reindividualization of these disorganized tissues might perhaps be analogous to reindividualization of the gemmules of freshwater sponges (Brøndsted, 1962), with the statocysts of Bryozoa and the reconstitution bodies of Ascidians. We are confronted with the same problem: how does an orderly new plan manifest itself during reconstitution? This is only a variant of the problem of how polarity is created in the young oocyte. This is therefore akin to the phenomenon of reindividualization and therefore a basic morphogenetic problem. I think that further cellular analysis of the reindividualization of the cysts in *Planaria velata* by histochemical and electron-microscopical techniques might be worthwhile.

Akin to the problems of heteromorphosis are those involved in the production of tumors, of great importance and just beginning to be studied so I will refrain from discussing them. Those interested should consult the papers of Seilern-Aspang (1960b, c), Stéphan (1962) and Lange (1966).

CHAPTER 15

INFLUENCE OF SIZE, AGE AND STARVATION ON REGENERATION RATE AND AMOUNT

IN Chapter 5, which dealt with the time-graded regeneration field, it was shown that the level of the incision in the body determined the rate of regeneration. If the rate affected planning an experiment, cuts had to be made at precise levels, but movements of the worms, or contractions under the influence of anesthetics, may result in alterations of the level of the incisions a little from individual to individual so to get valid figures for regeneration rates from specified levels, fifteen to twenty worms were necessary.

Child (1911b) strongly emphasized the need to use experimental animals in a uniform physiological condition. Several investigations have shown that age, state of starvation and size influenced the rate and also the kind of special regeneration; all these factors should therefore be taken into account when evaluating the results. By taking worms from their natural habitats and cutting them at a given level, there will be some differences in rates of regeneration in the various groups of specimens, even if animals of nearly the same size are selected; older underfed ones regenerate at a lower rate than younger well-fed ones, but it is difficult to distinguish such animals. Therefore to obtain dependable averages it is necessary to use animals of the same size, kept in the laboratory under uniform conditions, especially with respect to temperature and food it is preferable to grow clones from a single individual, either by fission, in species like *Dugesia dorotocephala, Phagocata vitta*, etc., or by cutting, in species as *Dugesia lugubris, Polycelis nigra*, etc., which do not propagate by fission.

After these more technical introductory remarks, let us look into the physiology of the phenomenon of the regeneration rate bearing in mind that the four parameters, size, age, state of starvation and temperature, may interfere with one another.

1. *Size of Fragment*

Child (1911b) stated that in *Dugesia dorotocephala* the rate and completeness of regeneration decreased with length of the regenerating piece and that this was not due to lack of material but to "physiological factors", but Abeloos (1930, 1932), in *Dugesia gonocephala*, found that the regeneration rate was independent of the size of the old parts and Buchanan (1933), in *Phago-*

cata gracilis, said that the "ability of head formation only slightly decreased in smaller pieces". Scharoff (1934) observed that small segments regenerate more quickly than big ones, but Rulon and Child (1937), using *Dugesia dorotocephala*, found a better regeneration of tail in head pieces when these were one-eighth of the body length than in those only one-sixteenth of the length. Chranowa (1939), in *Planaria Schischkowi*, stated that head regeneration was the same in animals with or without removal of the tail. Raven and Mighorst (1948) found that the rate of regeneration did not depend on the size of the piece in *Dugesia lugubris*.

The major question stated above has therefore been answered in conflicting ways by various authors. We know from the evidence discussed in Chapter 5 that the time-graded regeneration field is species-specific and therefore it is not an unreasonable inference that the rate of regeneration along the main axis of the animal is dependent on some factor or factors graded but strictly dependent upon the level in the healthy planarian body. This basic problem may be investigated by experimenting on segments removed from the rest of the body. If the rate of regeneration from a given region is the same in such fragments as that of the same region in unfragmented specimens, then we may be fairly sure that the time-graded regeneration field is a reality.

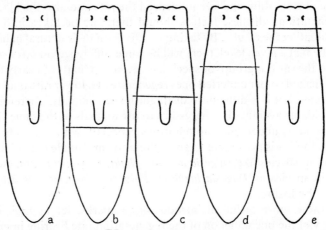

FIG. 125. *Dendrocoelum lacteum*. Explanation in text. (From A. and H. V. Brøndsted, 1954.)

A. and H. V. Brøndsted (1954) undertook a detailed study of the problem. Figure 125 shows experiments made on *Dendrocoelum lacteum* and Fig. 126 gives the time in hours of the head regeneration rate from the anterior surface of transverse cuts in five groups of animals, thirty in each group. It was found that the size of fragment did not influence the rate of regeneration, which started and terminated simultaneously in each.

Figure 127 shows an experiment on *Dugesia lugubris*. Eighty specimens were decapitated, forty as in *b*; both groups *a* and *b* regenerated eyes at the same rate, namely, after 116 hours. The sizes of eyes and regeneration blastemata were smaller in the *b* pieces, due to heavier demands on the neoblasts in these, which also had to regenerate tails from the posterior wound. The role of neoblasts in the rate of regeneration will be discussed presently.

Hours of regeneration	*Type of operation*				
	a	*b*	*c*	*d*	*e*
96	11	13	7	7	10
117	88	86	77	83	80
140	100	100	100	100	100

The types of operation are shown above. Thirty animals in each group.

Fig. 126. *Dendrocoelum lacteum.* Result of the experiment seen in Fig. 125. Thirty animals in each group. (From A. and H. V. Brøndsted, 1954.)

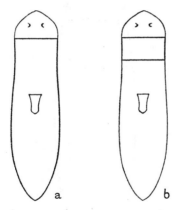

Fig. 127. *Dugesia lugubris.* Explanation in text. (From A. and H. V. Brøndsted, 1954).

In *Polycelis nigra* a similar experiment was made, only decapitation was made more posteriorly, just behind the eye-string (Fig. 128). Eighty specimens were operated upon, forty having a transverse segment $\frac{1}{2}$ mm long removed. The eyes began to appear on both sides at the base of the conical blastema (Fig. 129). After 145 hours in group *a* 68 % regenerated eyes whereas in group *b* 83 % did so. So in *P. nigra* the size of fragment is without influence on regeneration rate.

To see if heavier demands on available neoblasts influenced the regeneration rate from a given level in the time-graded regeneration field the experiment seen in Fig. 130 was conducted on *Bdellocephala punctata.* Sixty specimens were decapitated just behind the eyes; in thirty the right and left borders were removed exposing large regions from which regeneration could occur in addition to head regeneration from the anterior transverse wound. After 133 hours both groups *a* and *b* showed 33 % eye regeneration, after 154 hours, 70 % of group *a* and 65 % of group *b*, and after 179 hours, both groups reached 100 %; meanwhile the side wounds had formed blastemata. The experiment showed that extra demands on available neoblasts did not influence the rate of regeneration.

It is to be emphasized that the four species from four different genera exhibited the same basic features with the rate of head regeneration dependent

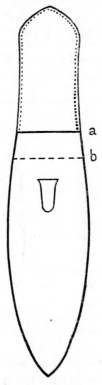

FIG. 128. *Polycelis nigra.* Explanation in text. (From A. and H.V. Brøndsted, 1954.)

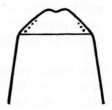

FIG. 129. *Polycelis nigra.* Explanation in text. (From A. and H.V. Brøndsted, 1954.)

only upon the level of a transverse incision in healthy animals. It is also relevant that none of the four species propagated by fission, it being reasonable to suspect that species propagating by fission might have other directing forces in their body, due to an accumulation of neoblasts near the fission plane.

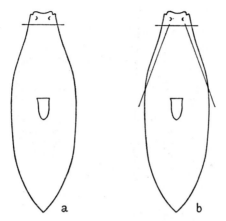

FIG. 130. *Bdellocephala punctata*. See text. (From A. and H. V. Brøndsted, 1954.)

The rate of regeneration expressed by differentiation of eyes—other organ formation has not yet been investigated—is dependent on factors located exactly in the time-graded regeneration field, but the later growth of differentiated eyes (head) is dependent on neoblasts available proved by Dubois (1949) in both irradiated (Chapter 7) and in healthy animals. When blastema formation is terminated the regenerating parts continue to grow until they reach a size conforming with the remnant of the body. A series of experiments performed by Brøndsted (1956) confirmed this. In *Dugesia lugubris* an experiment was made as in Fig. 127, and then extended as in Fig. 131. After regeneration of eye-spots in both groups the animals were fixed in formalin to obtain exact measurements. It was found in group *a* that the distance between the anterior tip of the blastema and the forerim of the pigmented part of the body (the old tissues) was $1·44 \pm 0·04$ mm; in group *b* it was only $0·96 \pm 0·04$ mm but it has to be borne in mind that group *b* had at the same time regenerated a tail (taxing the neoblasts), the blastema of which was $1·44 \pm 0·07$ mm. The eye-spots in group *a* had a longitudinal diameter of $23·12 \pm 0·59$ μ, in group *b* only $15·52 \pm 0·53$ μ. In the experiment (Fig. 130) on *Bdellocephala punctata* the longitudinal diameter of the eye-spots was, in group *a* $41·9 \pm 1·82$ μ and in group *b* only $35·22 \pm 2·11$ μ due to demands on the neoblasts on account of the lateral wounds.

The evidence from the above experiments was supported by Zeleny (1917), who found that regeneration was not retarded by simultaneous regeneration in other parts of the body.

The relation between the fixed time-graded regeneration field and its remoulding after regeneration was discussed in Chapter 5.

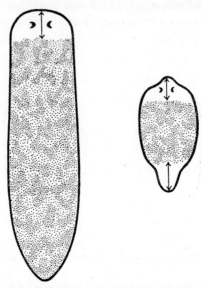

FIG. 131. *Dugesia lugubris*. Explanation in text. (From Brøndsted, 1956.)

2. *Influence of Age on Rate of Regeneration*

Child (1921) stated that physiologically younger animals exhibited better regeneration than older ones. This was confirmed by Abeloos (1927a and 1930), who found that both rate and quantity of regeneration was greater in younger animals than in older ones. Working with *Dugesia gonocephala* at 20 °C he found the rate of head regeneration to be the following: 3·5, 4, 5, 6 days in groups of 6, 11, 16 and 18 mm; the weight was 1·6, 6, 16, 22 mg. He also found that large animals regenerated better than small ones, rightly pointing out that the size of individuals was not indicative of age. Curtis and Schulze (1934) also found that large animals regenerated better.

In evaluating these statements it must be borne in mind that it is impossible to discern if the experimental animals were uniform in their other physiological parameters. Only experiments under strictly controlled conditions can give reliable results, therefore only specimens from clones raised under identical conditions can provide reliable answers. Temperature, food and crowding are the main conditions which must be taken into account. Individuals raised from cocoons may differ sufficiently genetically to give uncertain results. The problem is of considerable interest because the phenomenon called rejuvenation by Child is intrinsically involved, another of the basic biological enigmas. Most of those planarian species which can only propagate sexually seem in nature to have a more or less fixed individual life span of, say, one year. Planarians propagating by fission or fragmentation may be said to have infinite life. It is hard to say how an individual lifetime may be defined in such species, which makes them unsuitable for experiments aimed at elucidating our present problem involving age. Clonal individuals from species known to propagate exclusively sexually should give reliable answers, but again some doubts arise. If clones were formed by transecting the individuals into two or more transverse sections, then would the new individuals regenerated from these sections be of the same physiological age if anterior pieces regenerated to normality faster than the posterior ones? Can they still be said to be of the same age? How does this effect the rebuilding of the time-graded field as only when this is restored to normality is it possible to obtain reliable results as to the rate of regeneration. These difficulties might be circumvented by using clones consisting of individuals raised from specimens cut in half by strictly median incisions made at the same time in all individuals: after fixed periods one might then have clones of specimens of the same age. Only thus would it be possible to explore the question thoroughly. So far, I know of no such experiment, but I think that it certainly has to be made. Pending this, I think it better to omit a closer inspection of the problem of the influence of age upon the regeneration rate.

There remains the greater problem of possible rejuvenescence arising from regenerative processes. Child emphasized in several papers and in his books (1915b, 1941) that asexual reproduction and regeneration produce rejuvenescence. In 1911b he wrote: "Rejuvenescence occurs in asexual reproduction in Planaria", and in 1914b: "Senescence is associated with growth and differentiation and rejuvenescence with reduction and reconstruction."

The problem is basic and very intriguing. Growth rate in animals follows a curve rising steeply at first, thereafter diminishing, the curve ultimately being in most animals, especially warm-blooded ones, nearly asymptotic. Abeloos (1930) also showed this to be the case in *Dugesia geonocephala*. Child's opinion was mainly sound, and conformed with modern views on cellular behaviour that mitosis and differentiation were mutually exclusive processes.

We know that a planarian species propagating asexually gives rise theoretically to an unlimited number of individuals, and that a specimen from a species propagating only sexually may be forced to multiply indefinitely by cloning. In both instances it is clear that rejuvenescence takes place, but even if regenerative processes impose rejuvenescence on the animals this does not ndicate why young individuals regenerate faster than old ones.

In conclusion the following may be mentioned: in working out the time-graded field in *Bdellocephala punctata*, Brøndsted (1942a) showed that the field was already established in young worms recently emerged from their cocoons, and that the rate of eye-formation was the same as in adult (sexually mature) worms.

I think that the problem of possible higher regeneration rate in young animals postulated by several authors should be investigated in relation to the number of neoblasts available in the body; if regeneration takes place the number of undifferentiated neoblasts immediately after may be higher than in non-regenerated worms of the same size and age; hence only counting neoblasts can answer our question. The problem should also be investigated in relation to metabolic processes, e.g. neurosecretory substances like those found by Lender and Klein (1961), or possibly there may be higher metabolic rates in regenerating animals, a problem discussed in Chapter 17.

3. *Influence of Starvation on the Rate of Regeneration*

It is a well-known fact that planarians can endure prolonged starvation. Abeloos (1930) stated that adult *Dugesia gonocephala* worms starved down to a body length of under 1 mm resembled senile animals in that they were able neither to regenerate nor to take nourishment when food was presented and they died. Abeloos (1928), using *D. gonocephala*, showed that the size relationships between the organs during growth followed certain rules, the reverse occurring during starvation. The problem was therefore that if worms were deprived of food, how long could they retain a normal regeneration rate? The questions involved bear directly upon the whole problem of regeneration because an elucidation could provide valuable biochemical data as to the mechanism of regeneration especially as to the necessary materials.

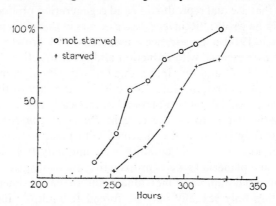

FIG. 132. *Phagocata vitta*. Explanation in text. (From Brøndsted, 1953.)

Brøndsted (1953) experimented on four species at a fixed temperature, $21 \pm 1\,°C$. Regeneration started from a transverse incision just behind the eyes in the anterior part of the time-graded regeneration field. In *Dugesia tigrina* head regeneration was delayed about 25% after 24 days' starvation. Well-fed *Phagocata vitta*, measuring about 6 mm when creeping, was only 2·4 mm when starved for 90 days, the body weight being reduced to about one-fifteenth of the normal. Figure 132 shows the retardation of the regeneration rate. *Dendrocoelum lacteum*, when starved for 16 days, lagged up to 20 hours in attaining the same regeneration level as the unstarved controls; when starved for 14 days, they only lagged 12 hours behind the controls. *Bdellocephala punctata* were starved for 34 days; Fig. 133 shows the result.

These results indicated that starvation for more than about 10 days retarded the regeneration rate, the effect differing in the different species. *Dendrocoelum lacteum* is a very active species when compared with the rather sluggish *Bdellocephala punctata*. It was therefore conceivable that a higher metabolic rate diminished the stock of metabolites necessary for regeneration in *Dendrocoelum* and retardation of the regeneration rate therefore occurred earlier in this species than in *Bdellocephala*.

It is a well-known fact that planarians starved for up to about a week regenerate at the same rate as fed ones, and are less prone to infections, supposedly due to the empty alimentary canal. In the different species the turning-point for declining regeneration rate seems to be 10–14 days, when the delay in regeneration rate may provide an opportunity to investigate which metabolites are most necessary and at what point in time. A beginning was made by A. and H. V. Brøndsted (1953), who showed that RNA accelerated regeneration in starved planarians (see also Chapter 18). In this connection the work of Willier, Hyman and Rifenburgh (1925) is relevant; the authors found that reserves of protein and fat were still unused after 6 weeks of starvation, and that fat was retained for 12 weeks. Abeloos (1930) was therefore right in saying that starvation was not a reason for rejuvenescence, as has been claimed by some authors.

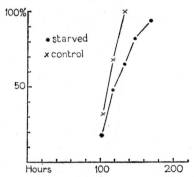

FIG. 133. *Bdellocephala punctata*. Explanation in text. (From Brøndsted, 1953.)

It is interesting that Mangebier and Jenkins (1964) found that succinoxidase activity in homogenates of *Dugesia dorotocephala* was greater in well-fed animals than in those starved for 48 hours. It is interesting because regeneration is not retarded after 48 hours starvation. Marino (1954) found the regeneration rate was retarded after starvation.

6 PR

CHAPTER 16

INFLUENCE OF TEMPERATURE
ON THE RATE OF REGENERATION

IT IS to be expected that temperature strongly influences the rate of regeneration but there are few exact data available. Abeloos (1927a), using *Dugesia
gonocephala,* found that at 20 °C the rate of head regeneration from three
levels of transverse incisions was 4½, 6 and 8 days, the former being the most
anterior and, at 14 °C the rates were 8, 12 and 14 days. Abeloos remarked that
these variations could be compared with head frequency. He proved (1927b)
that the regeneration of eyes from an anteriorly placed transverse cut at

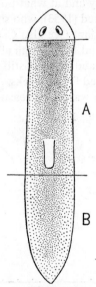

FIG. 134. *Dugesia polychroa.* Time-graded field. Explanation in text. (From A. and H. V.
Brøndsted, 1961.)

20 °C took 5 days, at 14 °C, 8 days and at 10 °C, 14–15 days. He wrote:
«Temperature intervient comme un simple catalyseur, pour modifier la
vitesse, mais non la nature intime, ni le résultat final de la tranformation.»
Abeloos (1929) found that in *Dugesia gonocephala* optimal growth was at

12 °C which suggests that growth and regeneration do not depend entirely on the same enzymatic processes.

The possible existence of other interesting features in regeneration rates detectable by temperature experiments stimulated A. and H. V. Brøndsted (1961) to investigate how two species from different genera would behave at various temperatures, the species differing considerably in the characteristics of their time-graded regeneration fields. *Dugesia polychroa* has a time-graded field almost identical with that of *Dugesia lugubris* (Fig. 27). On the other hand, *Bdellocephala* has a field illustrated in Fig. 20. The temperature-gradient chamber of Krogh was used, in which several compartments are separated by watertight metallic walls. Ice was put into the first chamber giving a series of compartments holding water from about 1·6 °C in the nearest to about 16 °C in the furthermost compartment, at a room temperature of 20–22 °C; the variation of temperature in the compartments did not exceed ±1 °C. *D. polychroa* was cut as indicated in Fig. 134. Sixty specimens were divided into six groups each containing A and B pieces; the controls regenerated at 20 °C. The other five groups were allowed to regenerate at 13 ± 1·0 °C, at 10·9 ± 0·9 °C, at 8·3 ± 0·7 °C and at 5·1 ± 0·5 °C. The duration of the experiment was 2 months. In Fig. 135 it is seen that lowering the temperature retarded the regeneration rate especially in the pieces from the posterior level in the time-graded field. *Bdellocephala punctata* was cut as indicated in Fig. 136. One hundred specimens were operated on and divided into five

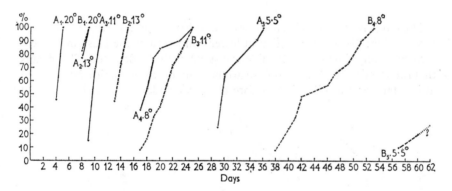

FIG. 135. *Dugesia polychroa.* Explanation in text. (From A. and H. V. Brøndsted, 1961.)

groups. The controls regenerated at 20 °C. The other four groups regenerated at 16 ± 1·0 °C, at 11·8 ± 1·0 °C, at 6·6 ± 0·6 °C and at 1·6 ± 0·4 °C. Figure 137 shows the results which basically agree with those of *D. polychroa*, although no regeneration occurred at 1·6 °C. Figure 138 and 139 show 50% in both species, but the regeneration rates of both A and B pieces were depressed differently by lowering the temperature in agreement with the findings of Abeloos. This feature has a bearing upon Child's hypothesis of a metabolic gradient along the main axis of the planarian body discussed in Chapters 5 and 22.

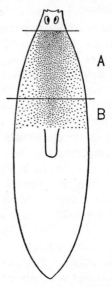

FIG. 136. *Bdellocephala punctata.* See text. (From A. and H. V. Brøndsted, 1961.)

The temperature experiments showed another feature, quoting the authors formulation: "It was found that at lower temperatures the blastemata were formed a long time before the first eye spots could be detected. This means that the wandering of the neoblasts to the wound was not seriously impeded at low temperatures." It is therefore to be concluded that the biochemical

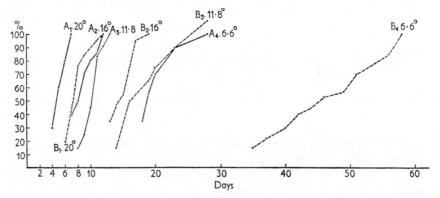

FIG. 137. *Bdellocephala punctata.* Explanation in text. (From A. and H. V. Brøndsted, 1961.)

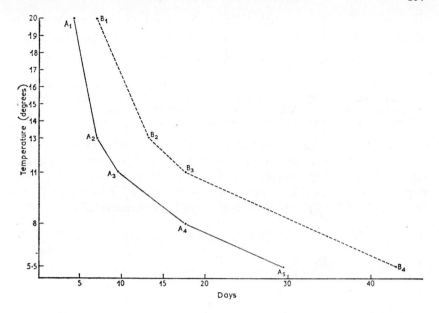

FIG. 138. *Dugesia polychroa.* Explanation in text. (From A. and H. V. Brøndsted, 1961.)

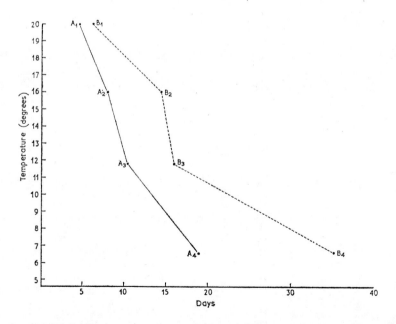

FIG. 139. *Bdellocephala punctata.* Explanation in text. (From A. and H. V. Brøndsted, 1961.)

processes necessary for cellular differentiation are more impeded by low temperature than migration of neoblasts. The authors remarked further:

> At low temperatures, the strong retardation or complete stoppage of processes of differentiation necessary for fulfilment of regeneration make the hypothesis of the adaptive value of regeneration doubtful: *Bdellocephala*, for example, has its egg-forming and egg-laying period from November to March, and the young are generally only hatched in April and May. So the *Bdellocephalas* have spent the better part of there lives at temperatures which impede regeneration.

This holds for Lake Furesø in the neighbourhood of Copenhagen.

Further temperature experiments may solve several problems of planarian regeneration, e.g. temperature coefficients for various enzymatic processes, migration rates of neoblasts, production of "organisin" (Lender), inhibition processes in various places of the time-graded regeneration field, and so on.

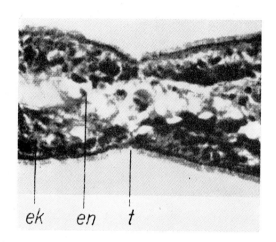

FIG. 140. *Microstomum*. Lateral sagittal cut. ek, ectoderm; en, endocytium; t, fission place (×400). (From Ax and Schulz, 1959.)

CHAPTER 17

ASEXUAL REPRODUCTION

SEVERAL species of planarians and other Turbellaria reproduce asexually either by simple fission or by forming cysts (Chapter 14), the former being the most frequent mode, often with more than one fission occurring more or less simultaneously.

Wagner (1890) distinguished between *architomy* and *paratomy*. The first denotes that the dividing animal splits transversely into pieces in which the new organs regenerate only after the fission process is terminated. The second mode denotes that several organs are regenerated before fission is completed. Ax and Schultz (1959) gave a short survey of fission in the Turbellaria. Architomy is the most common type of fission in the group Seriata, while the group Catenulida (e.g. *Stenostomum*) and the group Macrostomida (e.g. *Microstomum*) exhibit paratomy. In the group Acoela only a few species show fission, and of the paratomy type. The species used by Ax and Schultz could not be determined. The fission was by simple constriction without the formation of a septum, as is the case in *Microstomum* (Fig. 140). Both zooids formed a brain and statocyst before fission was completed. The authors gave a theoretical discussion of the phylogeny of asexual reproduction and concluded that paratomy was a phenomenon which arose independently in the different groups of the Turbellaria.

It must be pointed out that the authors still held that the tissues of the Acoela were syncytial but Pedersen (1964) recently has shown that the acoel *Convoluta* has cellular tissues.

Some authors may be mentioned from the older literature. Zacharias (1885, 1886) described fission in *Planaria subtentaculata* (= *Dugesia gonocephala, fide* Vandel, 1925) and *Polycelis cornuta*. After the first indication of constriction fission was completed in 3–4 days. The new posterior zooid formed a non-pigmented anterior outgrowth which developed into a new head, at the same time as the new pharynx, the epithelium of which was fully formed before the mouth opened, it is "... somit sicher mesodermalen Ursprungs". Zacharias found fission occurring until August, when the animals developed sexual organs.

Wagner (1890), in *Microstoma*, stated that fission started with the formation of a transverse septum of "connective tissue" followed by "Bildungszellen" (= neoblasts) accumulating behind the septum. In contrast to this

159

fission modus Ott (1892), in *Stenostomum leucops*, described the fission as a constriction without a transverse partition; the circular constriction deepening, transecting the nerve trunks and eventually the whole animal. Bergendahl (1892), in *Bipalium kewense*, discussed spontaneous fission and regeneration at considerable length, without producing many new facts. Voigt (1894) gave a useful survey of the then known species with asexual reproduction, and stressed the close resemblance to regenerative processes.

Child (1903a, b and 1904a) dealt meticulously with fission and regulative processes in *Stenostomum*. He maintained that the typical body form may be determined mechanically when the zooids were prevented from fixing themselves to the bottom of the dish by gentle water streams from a pipette. They required a longer time to attain the normal long and tapering form. Child therefore concluded that the normal body form was attained "... by the tension due to creeping and attachment...". He analysed both movements and muscle arrangements. Child (1904a) hinted at his well-known "dominance hypothesis" (Chapter 4) by saying that the oldest brain in the zooid chain dominates.

Both Ritter and Congdon (1900) and Curtis (1902) emphasized the close relationship between asexual reproduction and regeneration. Curtis found that fission was completed before organ formation took place and said: "hence there is produced the same appearance as in a specimen artificially cut in two at a definite point behind the pharynx". He found the existence of certain individual cells (neoblasts) to be responsible for the regeneration of new parts. Wilhelmi (1909) found a correlation between regenerative power and asexual reproduction, his views being of historical interest for their Lamarckism, as where the views of Colosi (1922, 1923).

Hartmann (1922), in *Stenostomum unicolor* and *S. leucops*, observed numerous agamous fissions for more than two years, and he consistently obtained the same result in fifty amputations.

Referring to more recent literature, Pedersen (1958), in *Phagocata vitta*, found fission to be of the architomic type;

> Just behind the pharynx a constriction forms, and an accumulation of neoblasts is observed at each side of the constriction. Mitoses are seen too. After some time the constriction breaks, a blastema forms at each broken end, and regeneration occurs, most extensively, of course, in the posterior part. Fission is very often released after decapitation, after a latent period of a few days.

Pedersen found that fission occurred most frequently postpharyngeally and only in a few cases prepharyngeally. It was emphasized that decapitation in this species also promoted fission; this promotion therefore seemed to be of a general phenomenon as claimed by Child in a long series of papers (1910c, 1911b, 1913d, 1914d, 1932, 1941). Child concluded that the removal of the cephalic ganglion meant the removal of the inhibitory influence on asexual reproduction, following which the physiological isolation of the second zooid was increased. Child (1911b) wrote: "The new zooid may be clearly defined dynamically before any evidence of morphological development can be discovered", and he suggested that the physical structure of the organism in which metabolism occurs may have some bearing on the subject, and that rejuvenation occurred during asexual reproduction in *Planaria*. He said (1914b): "Senescence is associated with growth and differentiation and rejuvenescence with reduction and reconstruction." This thought-provoking sentence, as said before (p. 151), conforms with modern views of cell division and determination as two processes excluding one another. Child (1932), using Japanese planarians, stated that when the heads were removed from bigger animals, fission was generally induced, and concluded that the head inhibited asexual reproduction. His "zooid conception" is profounded in his splendid book (1941). Van Cleave's (1929) experiments on *Stenostomum* seemed to support Child. In contrast Vandel (1921c) thought that asexual reproduction was a rhythmic phenomenon, fission being due to a simple mechanical extrication ("arrachment") and not to the formation of successive zooids, although he also stated that decapitation accelerated fission.

It is clear that Child's hypothesis is very interesting. We know from the investigations of Brøndsted (1942, 1954) and Lender (1955, 1956, 1960) that a head in the process of regeneration inhibits the formation of new heads, and that organs in general are inhibitory to the formation during regeneration of their like (Chapter 10). It is tempting to conclude that inhibitory substances from the brain are present in the whole intact body of the worm, although only in a slight concentration postpharyngeally. If this conception is right, then Child's notion of "dominance" of the head might be understood on a biochemical basis, and not, as Child emphasized, on the rate of metabolism, especially the rate of oxidation, as discussed in Chapter 17.

Perhaps the conception of inhibition by brain-inhibitory substances might somehow be connected with: (1) the different ability to reproduce by fission in different species; (2) the varying ability for fission in a given species under different external conditions; (3) the area of fission in the body when generally situated postpharyngeally. Let us consider these three questions to see if temptation leads us astray.

As to the first: one may think that in species with quick and easy head regeneration both inductive and inhibitory head substances should be present in high concentration in the body, so promoting rapid head regeneration and at the same time inhibiting head formation by fission. Species such as *Dugesia lugubris* can regenerate a head with great speed, but do not propagate by fission. On the other hand, species such as *Dugesia torva* and *Polycelis nigra* regenerate heads more slowly, but neither of them propagate by fission. On the other hand, several species of the genus *Dugesia*, e.g. *dorotocephala*, *tigrina* and others, propagate by fission and at the same time regenerate heads readily. So, concerning the first question, temptation is to be withstood.

As to the second, it is hardly possible to establish a relationship between amount of head-inhibitory substances and varying ability to perform fission under varying external influences.

The third question, the site of fission, may perhaps conform more with the conception of head-inhibitory substances graded posteriorly from the head, because prepharyngeal fission is rather unusual. The fact that fission occurs most frequently just behind the pharynx and not further back might be explained by the decrease in neoblasts caudally. If Lender's (1960) method of using brei or the supernatant from a centrifugated brei of heads were to inhibit fission, it might well be correct.

All this reasoning does not provide the clue for understanding how fission itself has evolved. The development of one or more new individuals and the breakdown of individual integrity still remain unsolved but very intriguing problems, basically the same as in all other animals and plants propagating by fission or budding. A discussion of this broad morphogenetic problem would lead us away from our main topic, planarian regeneration, but readers interested may refer to Child (1941), or Berrill's fine book (1961). For the time being we can only form working hypotheses as to the breaking down of individuality in planarians on the basis of experimental facts.

In the first place it must be emphasized that in many species asexual and sexual strains exist but there has been some dispute as to the respective classification. Kenk (1937) kept both sexual and asexual strains of *Dugesia dorotocephala* in the laboratory under identical conditions for 6 years without observing any change in the mode of reproduction which suggested that the propagating method was genetically determined. He then (1941) succeeded in producing gonads and copulatory organs in the body of the asexual form by removing the forepart of the worms and replacing it by grafts from the forepart of the sexual form. The experiment was not conclusive, however, because it was uncertain as to whether the gonads arose from the host or by a migration of neoblasts from the graft. Hyman (1939) apparently thought that external factors were decisive for the mode of reproduction. In *Dugesia dorotocephala* she found sexual forms in habitats where there was considerable movement of water, but asexual forms in stagnant water. Kenk (1940), however, kept sexual animals in the laboratory in still water, and they consistently reproduced sexually, with occasional fission in a few individuals. Vandel (1921c, 1925) held that fission was a rhythmic phenomenon caused by simple mechanical extrication and not to the formation of successive zooids as Child claimed. Vandel's conception of fission processes is therefore akin to fragmentation as seen in some species of the genus *Phagocata*. Vandel, in *Dugesia gonocephala*, stated that the asexual form seldom developed gonads. Castle (1927), in *Phagocata velata*, thought that a lowering of the temperature might induce sexuality. Benazzi (1942a,b) cultured clones of *Dugesia gonocephala* which propagated only asexually. In my laboratory a clone of *Phagocata vitta* reproduced asexually for 15 years at a constant temperature of 20 °C, with occasional specimens with traces of gonads.

Ogukawa (1957) and Ogukawa and Kawakatsu in a series of papers (1954, 1956a,b, 1957, 1958) studied the problem with various experimental methods. They found that sexual strains produced egg capsules from April to June, but from June to October *Dugesia tigrina* and *D. gonocephala* propagated by fission. The asexual strain underwent fission from March to December, but no sexual propagation was then observed; this strain seemed to be more hardy than the sexual one. It was of interest that the lowest temperatures for inducing fission in sexually mature animals was 17–20°; for sexually immature 13–16°, but for asexual strains 11–13°. In their natural habitats the life-cycle of the strains was very near to those in the laboratory. The authors concluded: "The periodical life cycle of maturity of gonads, development of sexual organs, breeding and fission of the animals seem to be dominated mainly by the seasonal change of environmental water temperature...".

Ogukawa and Kawakatsu (1956) found that several diets accelerated growth in *Dugesia gonocephala*, and they showed that the growth rate was positively and closely related to the frequency of fission in the asexual strains (for those interested in feeding planarians the paper cited should be consulted as it gives a wide range of diets). The diets did not seem to convert the asexual strains to sexuality, nor cause the sexual strains to become asexual. The authors further found that hormones from the pituitary did not induce asexual strains to develop gonads. Likewise Kenk (1941) did not find agamous strains developing gonads after feeding them with crushed tissues from sexually mature worms, a finding later confirmed by Ogukawa and Kawakatsu.

The same authors (1958) found frog Ringer and calcium-free solutions to be harmful to both sexual and asexual strains of *D. gonocephala*, even in concentrations as low as 1:16, confirming the results of Murray (1927, 1928). A pH of 4·0 depressed the fission rate, but the whole pH range from 5·0 to 9·0 was without harmful effect upon fission.

Phagocata vivida (Ijima and Kaburaki) is a planarian which propagates both sexually and asexually by a form of fission called fragmentation occurring in unfavourable conditions, i.e. above 20 °C, the animals simply constrict themselves in various parts of the body as seen in Fig. 141. The phenomenon has been studied by Kawakatsu (1959), who emphasized that fragmentation was entirely distinct from fission, which was a normal process in many species, a kind of regular budding, at prefixed levels, mostly postpharyngeally and proceeding rather slowly, whereas fragmentation occurred suddenly: and it seemed to be a phenomenon brought about as a direct result of the sudden contractions of the circular musculatures; moreover,

fragmentation occurred prepharyngeally as well as postpharyngeally. Kawakatsu stated that this phenomenon was known in only a few other species
of the genus *Phagocata*, and may perhaps not be a normal method of propagation but rather a pathological reaction. In *Phagocata vitta* the same

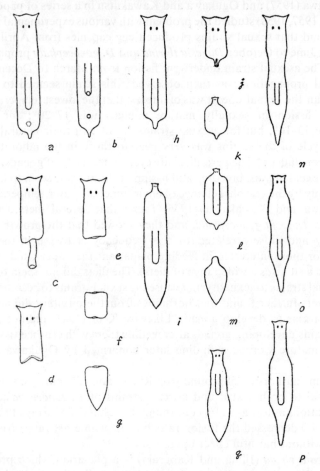

FIG. 141. *Phagocata vivida*. Various types of fragmentation. Semi-diagrammatical. See text.
(From Kawakatsu, 1959.)

phenomenon had often been observed in our clone in the laboratory, and
we think that it is a reaction similar to autotomy, but we have not investigated
the phenomenon very closely. Kawakatsu found that decapitation promoted
fragmentation.

Kanatani (1957a) followed up Child's (1932) investigations, in which
Dugesia gonocephala failed to undergo fission in the winter and early spring.

Kanatani kept the animals first at 3° to 9°C, when no fission occurred; an increase to 20°C initiated fission after 7 days. If the animals were decapitated fissions were more numerous. Kanatani then tried to see if aqueous extracts from the anterior portions of the worms had any influence on the frequency of fission, but these were without effect, except to cause production of supplementary eyes, perhaps due to Lender's organisin (Chapter 9).

Kanatani (1957b) continued his experiment on fission, using heparin, because of the well-known fact that heparin inhibits cell division. He kept decapitated animals in various concentrations of this (0·1–0·00001%) and found the curious fact that 0·1–0·05% heparin greatly retarded the frequency of fission, whereas lower concentrations greatly promoted it. Kanatani noted that heparin did not retard head regeneration indicating that neither the migration nor mitosis of neoblasts were inhibited. Kanatani suggested that heparin prevented gelation, which may be involved in the process of constriction; he also suggested that heparin, which is reported to inhibit phosphorylation in tissue cultures, was perhaps involved in ATP metabolism and ATP is necessary for tissue contraction. These observations seem valuable.

Kanatani (1957c) studied the effect of crowding on fission processes in *Dugesia gonocephala*. When ammonia excreted by the worms accumulated in the culture medium some supplementary eye formation occurred. Kanatani tentatively estimated that 2·2 μg of N/cc was the lowest concentration required to produce supernumerary eyes, but he also observed a marked increase in fission frequency in crowded animals and that the addition of ammonium chloride greatly reduced fission. He therefore concluded that the crowding resulted in the accumulation of ammonia in a high concentration and the inhibition of fission. If the culture medium were dialysed, the medium promoted fission. From this Kanatani tentatively suggested that a substance with a high molecular weight was liberated from the worms, and that this substance, being antagonistic to ammonia, was responsible for the higher fission frequency.

Kanatani (1960) treated *Dugesia gonocephala* with demecolcine (10^{-4} M) for 2–3 days at 20°, and colchicine. Sexual animals were thereby induced to undergo fission 3–4 times more often than the controls. The asexual strains were also influenced positively as to fission frequency. Kanatani and Flickinger (Chapter 6) showed that demecolcine considerably disturbed the polarity of the worms. Kanatani interpreted the phenomenon thus:

> … it is well known that the removal of heads will frequently induce the occurrence of fission and the results of the present study using sexual worms also showed this relation of the head to the occurrence of fission. On the other hand, it is known that the presence of a head seems somehow to inhibit the formation of another head at the posterior cut surface of very short pieces which would ordinarily become bipolar if the original head were not present (Child, 1941). It has also been shown by Flickinger (1959) that the presence of a head inhibits the formation of bipolar heads at the poste-

rior cut surface of transverse pieces when the original whole animals had previously been treated with demecolcine. Lender's (1956) results, which show that a water extract of the head prevents regeneration of the brain in *Polycelis nigra* and *D. lugubris*, suggest that such inhibition of head development is due to some chemical substance(s) whose concentration decreases from head to tail. All these facts indicate that the head plays an important role in maintaining the individuality in single axiate animals such as the planarian, and thus seems to uphold Child's theory of head dominance. In this connection it must be recalled that colchicine exerts a profound effect upon the nervous system (Ferguson, 1952; Eigsti and Dustin, 1955). Singer, Flinker and Sidman (1956) observed that colchicine destroyed the peripheral nerves in the regeneration of limbs in salamanders. Angevine (1957) also obtained similar results in mammalian nerve. Although the effect of demecolcine upon the nervous system has not yet been investigated, it is quite possible that this substance acts similarly to colchicine.

Pending further investigations we may formulate a few working hypotheses.

Individuality in animals is a phenomenon depending on a very delicate biochemical equilibrium between all components in the body. In some Phyla this equilibrium is unstable, well-known examples being several orders of Plathyhelminthes, Polychaeta and Oligochaeta in which there is a tendency for the body to separate into several individuals; here obviously the "zooid-theory" of Child (1941) must be taken into consideration. This theory, with its chief propositions of head-dominance and metabolic gradients, was dealt with in Chapters 4 and 17. The fact that decapitation promotes fission is, of course, strong evidence for "head-dominance"; and it is very tempting to ascribe this dominance to head-inhibiting substances; according to this hypothesis these substances would have to be regarded as the key to the preservation of individual integrity. Following this train of thought a clear relationship emerges between the existence of organ-inhibiting substances, the occurrence of neoblasts and regenerative ability.

When higher temperatures and better feeding in large worms promote fission, it could be due to a stimulation of the activity of the neoblasts, leading to a more active head formation in the postpharyngeal part of the body where head-inhibitory substances are apparently more scarce. Although it must be assumed that head-inducing substances are also scarce in this region, it is conceivable that the equilibrium between the antagonistic factors—inhibiting and inducing—are somewhat in favour of the latter, allowing the neoblasts to respond readily by differentiation into a head, and so develop a new individual. Following this train of thought must not exclude from our minds the problem of why other organs, e.g. the pharynx, fail to start a new individual.

When Kanatani (1957a) did not find an increase in fission rate on the administration of extracts from the anterior portions of the animals, this might have been due to a higher concentration of inhibiting substances from the head region. It is, however, still unsolved why exudates from crowded animals (Kanatani, 1957c) and demecolcine both promoted fission (Kanatani, 1960). Some possibilities present themselves. Kanatani proposed one and another is that demecolcine disturbs the equilibrium between inhibitory and inductive substances.

Even if these working hypotheses were followed up by experiments leading to a solution of the fission problem, there remains the curious fact that several species with high regenerative power and rich in neoblasts, e.g. *Dugesia lugubris*, do not propagate by fission. Why is the individuality of such species so rigid? They have plenty of neoblasts in the postpharyngeal region (Fig. 27) which respond readily to head-inducing substances when the worms are cut postpharyngeally, though enough head-inhibiting substances must be present to suppress postpharyngeal head formation in worms which are not decapitated. We might therefore imagine that the equilibrium of the two sets of substances is adjusted at a high concentration level. If this reasoning is acceptable it would seem that the ability to produce such substances is a species-specific character.

Fission and regeneration are thus closely interrelated and their problems are such that progress in research in the one is bound to cast light upon the other.

Still another problem remains: to what extent can fission and regeneration adapt to environmental stimuli and to what extent is the relationship between fission and sexuality an adaptive character?

CHAPTER 18

MORPHOGENETIC GRADIENTS AND LEVELS OF METABOLISM. CHILD'S HYPOTHESIS

CHILD's postulate (1910a, 1911c,d), proposed and modified over the years (1921, 1924, 1929a,b, 1941, 1946), is essentially that the rate of respiration is the basis of morphogenetic processes in planarians leading to orderly organ formation along the main axis, beginning at the head and progressing posteriorly. The head "dominates" due to its high level of metabolism expressed as rate of respiration. Child was confident that he had found a general formula in physiological terms, proposed as a purely quantitative entity with the level of respiration as the basic governing factor in all morphogenesis.

I regard this idea as wrong in principle. The rate of oxidation cannot, in my opinion, possibly be the cause of morphogenetic differentiation. The level of oxidation must depend upon, and be generated by, factors already located in the body, presumably bound to certain enzymic and/or hormonal constituents generated by RNA and controlled by the genetic code. Hence the basis of morphogenetic diversity must be fixed genetically. I am therefore convinced that the true initiator of morphogenesis in planarian regeneration is the reverse of that proposed: the level and kind of metabolism leading to morphological structures is ultimately caused by genetic factors, both in the blastema and in morphallaxis.

Child's hypothesis has, however, profoundly influenced the conceptions of morphogenesis, exemplified, for instance, in Huxley and de Beer's book (1934), and in summoning a wealth of other investigations. The evidence for and against must therefore be discussed at some length.

Child (1910a, 1911d, 1912b) thought that dilute ethanol and ether depressed oxygen uptake. The paper of 1910a is prophetic. He let *Dugesia dorotocephala,* his favorite experimental species, regenerate in 1·5% ethanol in 0·4–0·5% ether, and in 0·025–0·037% chloretone. He found that regeneration was slowed taking longer to form the right proportions between the various regions and organs. At that time he ascribed this to the diminishing movements of the animals (cf. p. 25). He also found that long pieces were necessary for full regeneration as many malformations occurred in short pieces, i.e. one to three eyes and sometimes more were formed, and, most significant of all, headless pieces were found. Child thought that all this, especially the headlessness, coincided with lowering of oxygen consumption and therefore concluded that lowered metabolism might be the overall cause for dysmorphic regeneration.

In (1911d) Child theorized at great length on the same problems and said that the use of dilute ethanol, etc., "... suggests that the relationship between anaesthetics and the organism is not wholly chemical in character, but may be connected, at least in part, with some characteristic change in the substratum i.e. the physical structure of the organism in which metabolism occurs". This somewhat cryptic comment recurred later without further clarification (1912b, 1921).

Child (1913b) combined his developing hypothesis with thoughts on "physiological resistance", that is, the length of life under the influence of KCN and anaesthetics; he emphasized that these substances influenced the rate of the "metabolic reactions or certain of them, probably oxidation". Later (1913c) he pursued this topic, again using KCN, and found that the disintegration of the worm with this respiratory poison started at the head and progressed posteriorly in stronger solutions. He simply postulated that the higher susceptibility leading to disintegration was due to the higher level of metabolism. He also enlarged his view as to the medio–lateral axes, and wrote: "The axial gradients in rate of reaction constitute the basis of polarity and symmetry in the organism, the region of highest rate dominating."

Child (1913d) clearly set out his hypothesis of dominance of the head region but a head to be formed before dominance appears. "The character of the head is also dependent on the rate of reaction in the head-forming region. As the rate decreases the heads formed become terato-ophthalmic, teratomorphic or anophthalmic, or finally the pieces may remain headless."

The head frequency may be altered experimentally by external factors which increase or decrease the rate of reaction, an idea which Child (1913d) summarized in the formula x/y, where x is rate of the head-forming region and y that of other parts.

In 1929b Child stressed that dominance and organizational activities started from the head and declined posteriorly in the first zooid of *Dugesia dorotocephala* (which propagates by fission behind the pharynx); polarity was therefore interpreted as an expression of dominance due to a quantitative and not a qualitative gradation; the head was the organizer and the quantitative gradation was due to the level of metabolism. He thus criticized both Morgan's view that polarity was dependent on structure, i.e. gradation of formative substances, and Runnström's view that morphogenetic gradients were due to different concentrations of certain substances, a view very like Morgan's. Child thought that his interpretation was correct because he and others had found graded susceptibility and permeability along the main axis of the planarian body to various poisons; even allowing for several sources of error in respiratory experiments, he still contended that the primary cause of axial-dependent susceptibility was a gradation of oxidation.

In his fine book (1941) he proposed a general hypothesis supported by a wealth of experimental findings in many different organisms. Those interested in the basic problems of morphogenesis must not miss reading this book.

In 1916 and again in 1930 Child used KCN in investigations of susceptibility and found that: "The longitudinal differential in susceptibility in the body wall of the first zooid *(Dugesia dorotocephala)* is the same in direction under all circumstances considered." He found the susceptibility varied in the transverse plane, a finding of note because it may indicate causes for susceptibility other than the level of metabolism, a question discussed presently. Child wrote that the varying susceptibility seemed to "indicate the existence of physiological differences in the parts concerned". Doubts as to the grade of susceptibility as a token for level of metabolism were clearly stated by Child (1932), who found that the extremity of the tail was very sensitive to poisons, and that differences in susceptibility between dorsal, ventral and lateral parts showed some degree of specificity for particular agents evidently a result of their different reaction in different direction.

In his comprehensive paper of 1946 Child wrote:

> ... that the axiate pattern in its beginnings and simpler forms consists in graded quantitative differentials in general metabolism, associated with graded enzyme and other differentials in their protoplasm. In general, intraorganismic reorganizers in reconstitutional development, such as the developing hydranth of *Corymorpha* or the developing planarian head, are regions of relatively high metabolic activity and exercise their reorganizing activity on lower gradient-levels or on regions without a definite axiate pattern.

And:

> ... indicate that the factors concerned are primarily quantitative differentials of some sort; ... are determined directly or indirectly by quantitative differentials in metabolism.

And:

> ... that quantitative metabolic differentials or gradients are fundamental factors of polarity and symmetry in their beginnings and their simplest forms.

What Child meant by "indirectly" is not clear, but it might perhaps be ascribed to certain misgivings about the fundamentals of his hypothesis.

Sivickis (1923), a pupil of Child, studied the same problems in *Planaria lata*. Here the superficial disintegration in KCN started anteriorly and posteriorly almost simultaneously; the last part to disintegrate was the region between the mouth and genital opening. Sivickis then meticuously described the disintegration rates of the various organs, maintaining that with proper precautions susceptibility towards KCN may be used as a measure of the level of respiration. His theorizing conformed with that of Child, but his paper was especially interesting because of the many experiments described.

Hinrichs (1924) studied the oxygen consumption in *Dugesia dorotocephala* using the Winckler method; his aim was to prove that caffeine depressed respiration but his figures were inconvincing. He stated that head frequency was stimulated by brief caffeine treatment but depressed by long application. He further stated that M/20 caffeine solution produced the type of disintegration usually seen by Child and others with several other substances. Later (1924b, 1926, 1927) he found that the axial gradient was also demonstrated by disintegration from ultraviolet and other forms of radiation, and he held that it coincided with the axis of metabolic activity in *Dugesia dorotocephala*. Anderson (1927), using the same species, studied oxygen consumption in alkaline water and found that it was accelerated; susceptibility was also augmented and this was ascribed to increased metabolism.

Buchanan supported Child's hypothesis by numerous experiments. He (1922) held that the transection of the worm enhanced respiration in two ways: (1) the wound itself increased the respiration of the cells nearby, and (2) the transection of nerves increased the respiration of the whole body. Weak anaesthetics were also found to increase oxygen consumption of excised pieces to such a degree that after 2 weeks it was many times higher than that of the controls. This claim will be discussed presently. Buchanan also held that head formation was determined by a tendency of the cells near the anterior wound to differentiate and to regenerate a head subsequently, and their capacity to inhibit head formation by cells in other parts of the body, a clear indication of the chemical inhibition processes discussed in Chapter 18.

Buchanan later (1923) reiterated and enlarged his views, saying that the stimulation at the wound was antagonistic to the overall stimulation of the whole body due to transection; if the former was high in proportion to the latter then a head was regenerated but if the latter inhibited more strongly then it should be possible to increase the head frequency again by the use of weak anaesthetics, which he (1922) claimed stimulated head regeneration. Using M/350 chloretone, a respiratory depressant, Buchanan showed that head frequency was increased, possibly because the metabolically more active anterior part of the animal recovered more rapidly from the anaesthetic than the older posterior tissues. There are many unknown parameters involved in these interesting experiments which require a critical analysis.

Using Winckler's method Buchanan (1926a) measured respiration in KCN and found it fell to about one-third of normal and then rose considerably above it. He thought that these experiments supported Child's susceptibility hypothesis. Later (1926b) he pursued his experiments and found that ether and ethanol gave a slight protection against the depressant action of

M/25,000 KCN on oxygen consumption. Buchanan (1930a) sought to sus-
tain Child's susceptibility gradient by placing worms in distilled water, where
disintegration is faster than in tapwater, and found it followed the antero–
posterior axis; this effect was thought to be of the same significance as the
respiratory gradient. Both disintegration and respiratory gradient were said
to be features of the differential of the antero–posterior axis in physiological
activity. It is suggestive that Buchanan (1930b) found that oxygen consumption did not rise appreciably in distilled water, and fell when disintegration
began; the poisoning effect was ascribed by Buchanan to excessive inhibition. The susceptibility gradient towards distilled water conformed closely
to the postulated respiratory gradient; worms in Ringer's solution (full or
half strength) remained intact for days, but when disintegration set in, it
started in the mouth region, so no evidence of an antero–posterior axis could
be found. Buchanan stated that Ringer's solution depressed oxygen consumption in the first hour, but later it increased it above the normal.

Buchanan (1934–5) later thought that the cytolytic antero–posterior gradient represented a differential gradient in susceptibility disturbances in the
calcium–lipid–water relationship in organs. The cytolytic gradient was generally parallel to the metabolic gradient and the death gradient, but not
necessarily so. Buchanan and Levengood (1939) concluded their investigations on the axial susceptibility by using antibodies. They injected planaria
brei into rabbits and the antibodies so produced were given to worms which
cytolysed in accordance with the usual antero–posterior gradient.

Strandskov (1934), using X-ray dosage of 172 to 864 r, found a susceptibility gradient conforming with those to chemicals. The same was found by
Gianferrari (1929) in *Dugesia polychroa* but *not* in *Polycelis nigra*. The head-
frequency curve is very much alike in these two species; but a susceptibility
gradient is not found in the latter casting doubt on the validity of Child's
concept. MacArthur (1921), using vital staining, and Miller (1931), using
strychnine, gave support to the susceptibility hypothesis. Watanabe (1937)
studied susceptibility in various concentrations of Murray–Ringer solutions
and found it conformed with the antero–posterior axis as it did to other
lethal chemical influences.

Before discussing the major problem of metabolic rate, claimed by Child
and others to be graded along the antero–posterior axis, some other causes
for susceptibility other than the level of oxidation must be pointed out. Wilson (1925, 1931), although strongly in support of Child's hypothesis, cast
considerable doubt on this conception in a meticulous description of the
cellular processes under the influence of KCN and dilute Ringer's solution.
M/1000 KCN in pond water locally invaded the sensitive lateral mucus-
producing regions. The rupture of these cells then permitted invasion by the
poison, so causing a dissolution of other cells. The disintegration was therefore not due to a specific action of KCN on the oxidative metabolism, but

to cytolysis, before the KCN had had time to depress oxidation elsewhere in the body. In 1931 the description was enlarged upon: "... the secretory granules swell prematurely between the cells of the hypodermis". This was followed by rupture of the dermis and a progressive disintegration due to the cytolytic action of the pond water. Diluted Ringer solution slowed down these cytolytic processes and he concluded that the axial gradient "... may be attributed to the reaction of the mucus-producing cells of the margin".

Perhaps the paper of Teshirogi and Maida (1956) should be mentioned here because it also dealt with the outer layers of the planarian body. The authors treated pigmented animals with thiourea and found that the pigmentation gradually disappeared from the head to the base of the pharynx, then from the tail forwards, so the pharyngeal zone was decolorized last. It is curious that the eye pigment was not affected by this procedure, but if the animals were cut into five pieces and treated with 0·1 % thiourea, then the eyes were gradually decolorized (in regenerants) more quickly in regions with high head frequency. When the pieces were put into water after the treatment pigmentation reappeared.

Wilson's findings and conclusions seem reasonable. The question therefore arises: may the features of susceptibility—which, as mentioned above, by Child and others, vary throughout the animal, from species to species, and do not even always conform with the main axis—be due, not to the level of oxidative metabolism, but simply to the different amounts of mucus-producing cells, either along the axis, the tail or the dorsal and ventral epidermis? With this doubt in mind we have to go into the basic question of whether oxidative metabolism really exists as a graded phenomenon along the antero–posterior axis.

1. *Direct Measurements*

Hyman (1919d, 1923, 1929, 1932b), using Winckler's method, found the following: the 1-pieces used 5·83 cm^3 oxygen per g in 24 hours, the 2-pieces, 5·19 and the 3-pieces, 5·78. The 2-pieces shortly after the transection gave a higher quotient than the 1-pieces but the differences were exceedingly small: 1·03 and 1·05 cm^3 oxygen per g in 4 hours. It must be stated that these differences were too small to allow the conclusion that an oxidative gradient existed along the main axis. The Winckler method is too inexact as is weighing small pieces, due to water content and other reasons.

Child (1919) supported Hyman's findings (1919d) using a colorimetric method with phenolsulphonephthalein. He also confirmed that KCN inhibited respiration measured as diminished carbon dioxide production. Parker (1929), Shearer (1930) and Needham (1931, 1942) criticized Hyman's results, especially because the Winckler method was regarded as inadequate in the instance given. Parker and Shearer resorted to the Warburg method, Shearer using nitrogen determination by Kjeldahl after respiration determinations. Neither of them found any evidence of a metabolic axial gradient. Shearer wrote: "The outstanding feature of Hyman's work on the respiration of planarians is the remarkable uniformity of all her results." Hyman criticized Parker (1932b) and firmly held to her opinion throughout a lengthy discussion. She was supported by Watanabe (1935a) and Child (1941). Lewetzow (1939) did not think that an antero–posterior respiration gradient was present in Polyclads, although using the Warburg method, he found slightly higher respiration in the foreparts of the animals than in the posterior parts, but the scatter of the determinations was wide. Løvtrup (1953), in my laboratory, reinvestigated this controversial subject using *Dugesia lugubris*. In order to avoid the errors of Winkler's method and damage to the worms by shaking in the Warburg apparatus, Løvtrup used the Cartesian diver micro-respirometer of Linderstrøm–Lang and Holter (1943); moreover, instead of the wet weight, which is an unsatisfactory reference, she used the total nitrogen content as standard. Løvstrup found that the rate of respiration was 10 μl per hour per total of nitrogen; she did not find any respiratory gradient conforming with the antero–posterior axis.

Flickinger and Blount (1957) in an otherwise stimulating paper could not find any decline in respiration from head to tail in *Dugesia dorotocephala*. They used an ingenious method based upon measurements of the e.m.f., which did not reveal the actual amount of oxygen used in the metabolic processes, but the relative rate of respiration; for the planarians used this was found to be: head 10, middle 10, tail 10. The authors took the precaution of immobilizing the animals in 1% agar thus circumventing the necessity for cutting the animals into pieces, cutting having previously aroused criticisms, because it might influence the level of respiration. The authors were aware of this possibility, but their results nevertheless confirmed those of Løvtrup.

Two features are pertinent to our problem: does wounding influence respiration and starvation influence it further subsequently?

Fraps (1930), using a modified Warburg method, claimed: "... the respiration of Planaria is at a maximum immediately following sectioning". It is not quite clear if this postulate is fully substantiated but apart from this his paper is not relevant to Child's hypothesis.

Pedersen (1956), in my laboratory, undertook a meticulous study of the respiration of *Phagocata vitta*, a species propagating by fission. The animals were genetically pure, being cloned from one specimen. He used the diver method modified by Zeuthen (1950), the micro-Kjeldahl (Hawk, Oser and Summerson, 1947) a modified Levy (1936) and Doyle and Omota's methods (1950). Intact animals, starved for 2–12 days in order to evacuate the digestive system, had a respiratory rate of $8 \cdot 4 \pm 0 \cdot 3$ μl oxygen per mg total N per hour. Decapitated animals (= wounded 1–2 hours after the operation) showed a respiratory rate of $8 \cdot 6 \pm 0 \cdot 4$ μl oxygen, and 24 hours later, $8 \cdot 0 \pm 0 \cdot 2$ μl. During fission the rate was: $7 \cdot 9 \pm 0 \cdot 4$ μl. These figures showed that neither wounding nor fission had a significant influence upon the respiratory rate. Nor did Løvtrup find an increase in oxygen consumption after decapitation (wounding). Pedersen (1956) wrote: "Only more extensive injury such as cutting into small pieces cause an increase in the respiratory rate. It seems natural to refer this increase to processes of cytolysis." Hence it seemed very probable that the increase in respiration postulated by various authors was small in clean and slightly damaged animals, and may have been due to the unavoidable cytolysis taking place in such small injuries.

The second feature, the influence of starvation on the respiratory rate, has also been studied by Pedersen (1956). As discussed in Chapter 15, the unreliability of using animals of unknown age and feeding state in physiological experiments caused Pedersen to use cloned *Phagocata vitta* of the same age and feeding state and he took the further precaution of making sure that no cannibalism took place. Using the same techniques as in his decapitation experiments he found that the oxygen consumption was unaltered during starvation for 35–38 and for 63–66 days.

The method of susceptibility to KCN and other supposed depressors of respiration was shown to be inadequate for evaluating respiratory gradients along the axis, as were direct measurements of oxygen consumption. These results with modern methods disproved the idea of an oxidative gradient in conformity with the morphological one in planarians. The indirect method of Flickinger and Blount also disproved the existence of an oxidative gradient, as well as several other experiments carried out with reductive dyes and LiCl aimed at proving or disproving Child's hypothesis.

Robbin and Child (1920) used, as said above, phenolsulphonephthalein as a colorimetric indicator for CO_2 production. Here again weighing was liable to be a serious source of error, as was motor activity, although the authors found it nearly absent at 18 °C. They found CO_2 production ran parallel with changes susceptibility. The CO_2 production (in *Dugesia dorotocephala*) was markedly increased in pieces cut from near the muscular mouth opening but only slightly if at all increased in pieces from regions near the head. Development of a new asexually propagated individual meant a considerable increase in CO_2 production, which involved the old parts as well (one must again assume muscle contractions to be the cause).

Brøndsted (1937) used the unpigmented *Dendrocoelum lacteum* in order to observe clearly the colour changes of methylene blue, which enters oxidative processes as a hydrogen acceptor, being reduced to the leuco compound under anaerobic conditions. He used this species also on account of its steep head-frequency gradient ceasing abruptly just anterior to the pharynx (Fig. 103) and therefore it is well suited for testing Child's hypothesis. The animals were kept for 16–17 hours in 0·0001 % methylene blue in tap water, rinsed and then immobilized in 1/300 chloretone in Carrel flasks. Nitrogen was passed through the flasks for 5 minutes. The colour was unevenly distributed from the start. Figure 142 shows the regions mapped out on the animals. The colours of these regions were compared with those of twelve strengths of methylene blue on white filter paper dipped in the dye. The glands and nervous system were strongly coloured and the parenchyma only feebly, as shown in Fig. 143. The incline of the curve indicates the rate of decolorization. Figure 144 shows the colour intensities of the ten pieces of the animals

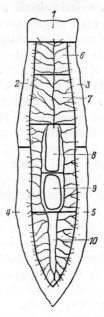

Fig. 142. *Dendrocoelum lacteum*. See text. (From Brøndsted, 1937.)

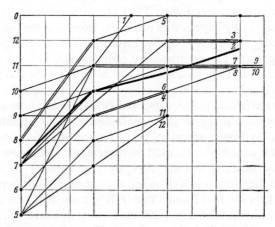

Fig. 143. *Dendrocoelum lacteum*. See text. (From Brøndsted, 1937.)

after $\frac{1}{2}$ and $1\frac{1}{2}$ hours in relation to the initial colour intensities. It can be seen that the oxidative capacity increased from the head to the gut, whence it declined posteriorly. The experiment therefore did not support Child's hypothesis, but seemed rather to indicate the oxidative capacities of the "daily" physiological work of various parts of the animal.

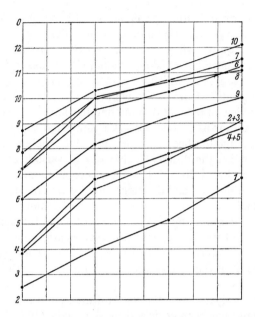

FIG. 144. *Dendrocoelum lacteum.* See text. (From Brøndsted, 1937.)

Child (1948), in *Dugesia dorotocephala,* found that Janus green was reduced first on the ventral surface of the animal, the rate decreasing posteriorly in the first zooid; in *D. tigrina* the sides and ventral parts were oxidized the first; the head end reducing the dye at the highest rate with reduction progressing more slowly posteriorly with a slight gradient. Methylene blue did not show such an overall graded direction, but the cephalic ganglia reduced the dye most quickly, the reduction progressing postero-pedally along the nerve cords and the regions between them. Child (1949b) in *Stenostomum* obtained similar results with both dyes.

Using *Polycelis sapporo* Watanabe (1948) showed that the regeneration rate for head and tail formation was quantitatively related to the level of the body following the antero–posterior differential. Watanabe (1955) set out to provide "... evidence for the presence of an axial gradient or gradients in an unpigmented single-zooid triclad, *P. sapporo,* by making visible the differentials of susceptibility to ethyl alcohol, sodium cyanide, the sulphates of copper, cobalt and cadmium, and the oxidation and reduction of methylene blue". He confirmed the older results of Child's school as to an antero–posterior disintegration axis in high concentrations of the chemicals, but the reverse in low concentrations of ethanol and NaCN. The many interesting features of the behaviour of heteromorphoses should be looked up in the original paper. Watanabe also used methylene blue to prove the existence of an antero–posterior oxidation gradient. In this section of his paper he criticized Brøndsted (1937) as he thought that 16–17 hours in the concentration used was "... quite sufficient to alter or even obliterate whatever gradient patterns might have been visible originally". He also criticized the mild anaesthetic chloretone: "... which is also a very good agent for decreasing or obliterating gradients". I am quite willing to accept Watanabe's critical remarks, even in view of the fact that my findings have been accepted by such an experienced author as Brachet (1947). I think that the problem deserves a rigorous reinvestigation because other different problems could be solved at the same time, i.e. the different behaviour of various cell types in their use of methylene blue. I do not, however, accept Watanabe's opinion that planarian susceptibility to, and disintegration by certain chemicals are proof of a graded axial oxidation corresponding to the morphological antero–posterior axis, which in many cases agrees with head frequency, i.e. rate of head regeneration. Even if colorimetric experiments should prove the existence of an antero–posterior oxidative gradient I am unwilling to concede that this necessarily means that such a gradient is the cause of the declining head frequency in some species. Here I am in agreement with several other authors, i.e. Bertalanffy (1941) and Bertalanffy, Hoffmann and Schreier (1946) even though Bertalanffy has been criticized by Watanabe (1955).

I am unwilling to accept the doctrine of Child, his pupils and co-workers, because I consider that the papers of Løvtrup (1953), Pedersen (1956) and Flickinger and Blount (1957) disprove tha existence of an antero–posterior oxidation gradient.

Pending further experiments aimed at elucidating the cellular mechanisms responsible for the graded susceptibility and possible gradation of oxidation of certain dyes determined colorimetrically, I think it safe to seek other causes for the existence of an antero–posterior head-frequency gradation found by Child, and a time-graded regeneration field, found by Brøndsted.

Flickinger (1963a) defended Child in a stimulating paper by saying:

> Much of Child's work was devoted to the demonstration of respiratory gradients in various organisms and embryos, and perhaps because of this it is often thought that Child's metabolic gradient concept referred only to axial differences in respiration... However, the essence of Child's theory is that axial quantitative differences in metabolic rate, not merely oxygen consumption, can direct axial polarity in development.

And later: "... Child's gradient theory should be examined in terms of protein syntheses or turnover, instead of oxygen consumption". Flickinger proved the existence of such a gradient in protein turnover. This will be discussed presently.

Before doing so, however, a few remarks should be made regarding Flickinger's defence of Child's hypothesis. There can be no doubt as to Child's and his co-workers' view that the amount of metabolism, that is, the overall rate of metabolism is the underlying cause of axial patterns. With such tools as were available, Child was unable to unravel which specific types of metabolism were involved and which directing. He and his followers therefore had to cling to a hypothesis of an overall oxidation rate. Flickinger himself was unable to prove the existence of such an oxidative axis. The level of oxidation was not the cause of an axial pattern as meant by Child, but this does not in the least imply that some dormant metabolic influence might exist in the planarian body directing head frequency and other polarity phenomena. On the contrary, variations are bound to exist. Perhaps the inducing capacity of Lender's "organisins" provides an instance of such a variation, which, when the animal is wounded, leads to the stimulation of a certain type of metabolic

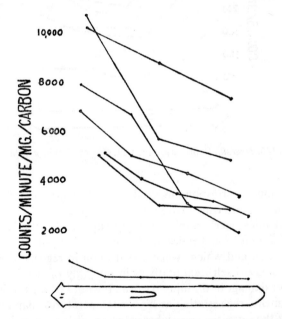

FIG. 145. *Dugesia*. Explanation in text. (From Flickinger, 1963a.)

pattern. If this reasoning is sound the axial pattern is morphologically and biochemically the basic cause for regeneration phenomena, and the result is a specific rate of oxidation. We shall see presently that Wolff and co-workers gave another explanation for Child's gradient hypothesis. Although, in my opinion, Flickinger's defence of Child was wrongly based, his actual findings are of the greatest interest.

Flickinger determined the specific activity of the protein fraction of worms after incubation with $^{14}CO_2$. Figure 145 gives his results. It may be seen that the protein turnover closely follows the axial gradient as expressed by head

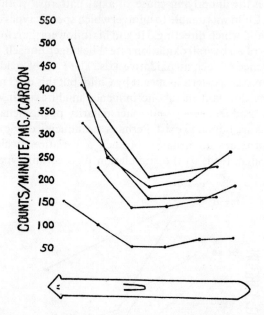

FIG. 146. *Dugesia*. Explanation see text. (From Flickinger, 1963a.)

frequency without a corresponding gradation of respiration rate. Figure 146 shows the situation after incubation with ^{14}C-glycine.

I think that Flickinger's results strongly support Morgan's view (p. 25) that some "dormant" biochemical design lies behind polarity, which is of a qualitative nature, and which, when called upon in regeneration, goes into action and displays itself quantitatively in one way or another, depending upon the site of damage in the injured worm. It would be a good idea to see if posterior segments regenerating at higher temperatures than anterior ones grew heads at the same or even higher rates. In such an experiment optimal temperatures for anterior head regeneration would have to be taken into consideration.

Recently Wolff, Lender and Ziller-Sengel (1964a) proposed an ingenious plan for the existence of physiological gradients in regeneration in planarians by saying: "The theory of axial gradients (Child) and the concept of dominance can be explained by the presence of specific inhibitory substances which

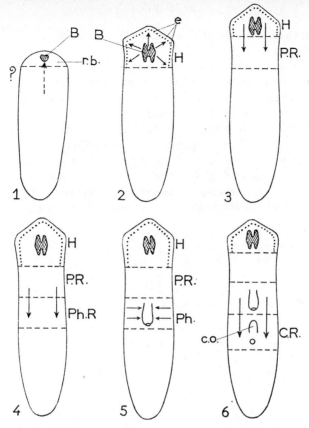

FIG. 147. *Polycelis nigra*. See text. (From Wolff, Lender and Ziller-Sengel, 1964.)

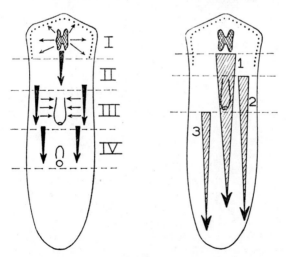

FIG. 148. *Polycelis nigra*. Diagram of inductions, left, and inhibitions, right, presumed by the Wolffian school. See text. (From Wolff, Lender and Ziller-Sengel, 1964.)

diffuse from a centre where their concentration is maximal, to distinct regions where their concentration is minimal."

The inhibitory phenomena were discussed in Chapter 10. Here it is sufficient to present the schematic Figs. 147 and 148, where the inductive processes complementary to the inhibitory ones are also indicated. In Chapter 23 we shall return to the whole problem in an attempt to co-ordinate the facts with regard to head frequency and the time-graded regeneration field along with other basic questions in planarian regeneration.

Let me close this chapter with the remark that I have refrained from a thorough discussion of Child's hypothesis covering all phyla and classes in the animal kingdom, because I think the basic circumstances vary exceedingly, and such a discussion would lead far astray from the aim of this book.

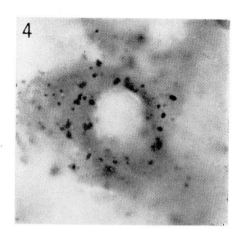

FIG. 149. *Dugesia lugubris*. Neoblasts incorporated with uridine-H[3]. See text. (From Urbani and Cecere, 1964.)

CHAPTER 19

BIOCHEMICAL INVESTIGATIONS
ON PLANARIAN REGENERATION

IT MAY be considered established that neoblasts build the blastema and are rich in RNA, which is mainly distributed in the sparse cytoplasm as "free" ribosomes (Pedersen, 1959b) but also attached to the barely present endoplasmic reticulum. This may mean that the RNA is not yet organized to perform specific functions in the neoblast, or that the cytoplasmic RNA in these "embryonic" cells is not yet completely controlled by the chromosomal DNA. If this reasoning is sound, the chromosomal DNA should go into action only when the blastema is forming, and the neoblasts then begin to be determined and later differentiated by the chromosomal DNA, but always in conjunction with stimuli from adult tissues. There seems to be a feed-back system (Chapter 22) from adult tissues acting on chromosomal DNA, forcing the genes to organize the RNA in the neoblasts to manufacture the proteins specific for the particular cells regenerating missing tissues. Brachet (1960b) has shown that the RNA in neoblasts is involved in the synthesis of proteins.

Urbani and Cecere (1964) in *Dugesia lugubris* showed that tritium-labelled uridine was incorporated after 24–28 hours mainly in the cytoplasm of the neoblasts (Fig. 149). This may be of great significance in understanding the dynamics of neoblasts. In Chapter 22 we shall examine this basic problem again. Meanwhile we must look into the biochemical evidence produced by many authors.

Urbani (1955, 1962) and Urbani, Bellini and Zappanico (1958), working with *Polycelis nigra*, showed that in the regenerative zone of the tail proteinases reached a maximal activity after the first day, the dipeptidases after 3–4 days, and after 5 days, that is when the regeneration was nearly complete, both enzymes returned to normal. These studies were confirmed and extended by Bellini, Cecere and Zappanico (1962) using tail regeneration in *Dugesia lugubris*. It was mentioned (p. 8 and 80) that a zone of liquefaction developed at the wound beneath the newly formed epithelium, due to some cytolysis. It is significant and interesting that Bellini *et al.* (1962) found that proteolytic activity reached a maximum at 4 days (Figs. 150, 151, 152) whilst that of dipeptidases was at 6–8 days; the maximal rates of incorporation of ^{14}C-methionine and ^{14}C-leucine were attained in the blastema about 6 days after cutting. It therefore seems that proteinases may be responsible for the

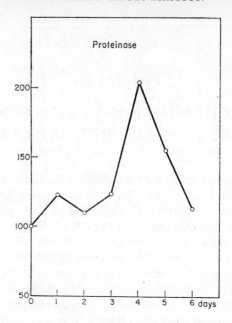

FIG. 150. *Dugesia lugubris*. Explanation in text. (From Bellini, Cecere and Zappanico, 1962.)

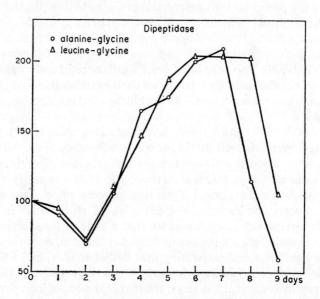

FIG. 151. *Dugesia lugubris*. Explanation in text. (From Bellini, Cecere and Zappanico, 1962.)

liquefaction of the damaged tissues before the rebuilding of proteins after the action of the dipeptidases. These findings are only provisional in the clarification of cell-determination, but they are very suggestive, and may perhaps lay the foundations for further investigation along these promising lines,

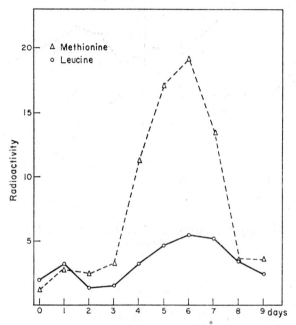

Fig. 152. *Dugesia lugubris*. Explanation in text. (From Bellini, Cecere and Zappanico, 1962.)

especially for the study of neoblasts by EM and histochemically, at times corresponding to the incorporation of ^{14}C-labelled amino-acids. Urbani (1962b, 1964, 1965) in stimulating papers has discussed the general problems involved in morphogenesis. This was followed up by Autuori, Nardelli and Gabriel (1965) who found (Figs. 153, 154) that the activity of proteinases and acid phosphatases increased during the first 48 hours of regeneration in *D. gonocephala*; they rightly thought that these enzymes degraded and liberated substances, which were later used for regeneration. This suggestion was endorsed by their finding that dipeptidases and alkaline phosphatases increased further after 3–4 days of regeneration, which they interpreted as leading to active synthesis of the specific proteins necessary for differentiation. They found that this activity was higher in well-fed animals than in those starved for 5–6 weeks; this is interesting in relation to the rate of regeneration of starved animals discussed in Chapter 15.

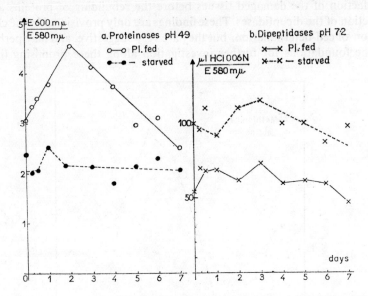

FIG. 153. *Dugesia gonocephala.* Explanation in text. (From Autuori, Nardelli and Gabriel, 1965.)

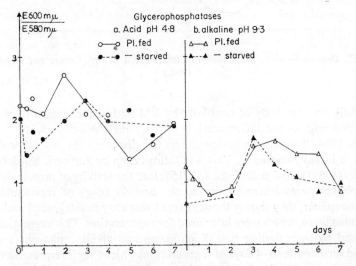

FIG. 154. *Dugesia gonocephala.* Explanation in text. (From Autuori, Nardelli and Gabriel, 1965.)

Osborne and Miller, Jr. (1962, 1964), using *Dugesia tigrina,* started from the hypothesis that an analogy exists between lysosomes and food vacuoles, at least in lower animals (Jennings, 1962), and that lysosomic enzymes are involved in the autolytic degradation of tissues in metamorphosis and involution; from this idea they thought it likely "that acid hydrolases may play a similar role in the gradual disappearance of digestive and reproductive organs during prolonged starvation in planarians, as well as in the provision of raw material for the early stages of regeneration following transection, before the ingestion of exogenous foodstuffs can be resumed". The authors, using biochemical methods, found that resting neoblasts did not show acid phosphatase activity, but that in the regeneration zone they showed much alkaline phosphatase activity. They wrote: "the significance of this observation is uncertain since alkaline phosphatase activity also characterizes resting neoblasts. ... It is suggested that the lysosomal acid and hydrolases (typified by acid phosphatases) are involved not only in the early stages of digestion in the food vacuoles, but also in the autolysis of dispensable organs during starvation and in the tissue breakdown which precedes regeneration in transected animals."

Reflecting on the findings quoted above, it seems that the first biochemical steps in regeneration are beginning to be understood.

Brachet (1960a) suggested that morphogenesis was controlled by a system involving a sulphydryl-disulphide equilibrium; Descotils *et al.* (1961) studied several types of animals, including planarians, with this in view; the regerating animals were cultivated in β-mercaptoethanol and dithiodiglycol solutions of various concentrations; the regeneration processes were studied by means of cytochemical and autoradiographic methods. Mercaptoethanol appeared to inhibit regeneration, whereas dithiodiglycol had almost no effect. Mercaptoethanol increased to some degree the basophilia of the regenerating tissues and stimulated RNA synthesis. The authors thought that their results confirmed the importance of —SH groups in morphogenetic movements, a link perhaps with the earlier findings cited above.

It is of note that Pedersen (1959) and Lender and Gabriel (1960) were unable to demonstrate alkaline phosphatase in neoblasts, but that —SH groups were present in them both before and during regeneration.

The work of Gabriel (1963) showed that in *Dugesia gonocephala* β-mercaptoethanol did not inhibit the migration of neoblasts, which although still rich in RNA, became pycnotic and failed to differentiate. He was unable to say whether β-mercaptoethanol influenced the SS–SH equilibrium, directly on the enzymes, or on induction and/or on the inhibitory substances.

Now we must not concentrate only on neoblasts; it is more than probable that other tissue cells are involved in the initial regeneration phases. Gaszó, Török and Rappay (1961) found alkaline phosphatases in the nervous system of *Dugesia lugubris* confined to the nuclear membrane, nerve fibrils and the neuromuscular connections; specific cholinesterase was present but only in somatic neuromuscular junctions in motor neurones. It was curious that they did not find similar positive reactions in *Dendrocoelum lacteum*; they suggested that this might have been due to the lesser regenerative ability in this species. Against this explanation, however, I must say that in my experience the regenerative ability of *D. lacteum* is very good, except that no head regeneration occurs behind the pharynx, however, head regeneration from the forepart and tail regeneration from the whole body occurs readily. The difference in the amount of the enzymes in the two species investigated by the authors might be explained by the much greater motility of *D. lacteum*, so that they use up their enzymes at a greater rate. Whatever the explanation it is very interesting that Ichikawa and Ishii (1961), in *Dendrocoelopsis lacteus*, found that the nerve cords after transection grew thicker and looser in texture, apparently assisted by neoblasts. In the intact animal they found alkaline phosphatase in the brain, pharynx and penis papilla, with a high activity of this enzyme in the nerve cords as well. The authors thought that the alkaline phosphatase present during regeneration came from the brain, possibly a biochemical link with the findings of Urbani and the other authors mentioned above.

In (1955) I wrote: "Investigations with modern techniques are sorely needed and would surely repay efforts." Such investigations have since been undertaken by several authors, especially by Flickinger and his co-workers, in a stimulating and provocative manner. I therefore intend to review and discuss this work at some length but beginning with some early investigations. Lecamp (1942) found that amino-acids in a concentration of M/100,000 in water accelerated regeneration, with increasing effect in the order: glutamic acid, glycine, cystine, lysine, tryptophane, arginine, histidine; a mixture of tryptophane, arginine and histidine had the greatest accelerating effect. Tryptophane was not found to be a factor for differentiation, but the eyes were a little slower in appearing than in the controls. Lecamp investigated at a constant pH the effect of glycocol, *d*-glutamic acid (both basic and as chlorhydrate), *l*-asparagine, *l*-leucine, *d*-lysine (bichlorhydrate), *l*-histidine (monochlorhydrate), *l*-cystine, *l*-tryptophane, amino-alcohol, choline (in concentrations of M/1000 to M/100,000). In concentrations of M/5000 lysine, cystine and arginine were poisonous. From M/5000 to M/40,000 regeneration was slower than in the controls. The length of the animals was used as a measure of regeneration rate, but this was not a very reliable criterion. It was proposed by Brachet (1950b) that the basic amino-acids arginine and histidine were perhaps necessary for the synthesis of purines in the nucleic acids. The results of Lecamp were only preliminary and full of unanswered questions, such as: do the specified amino-acids assist in accelerating blastema formation by promoting the migration of neoblasts, or take part in forming proteins; do they simply act as nutriments or take part in nucleic acid metabolism and so on? Some facts have been established: Clement (1944) found that the basophilia of cytoplasm and nucleolus of the neoblasts was due to RNA and it disappeared during organogenesis. A. and H.V. Brøndsted (1953) promoted regeneration in starved planarians with RNA, and found that the neoblasts used up their store of RNA in protein synthesis in the blastema.

Coldwater (1930, 1933) and Goldsmith (1934) had already suggested that the liberation of sulphydryl groups might be essential for regenerative processes, an idea extensively discussed by J. Needham (1942), Brachet (1950b), A. E. Needham (1952) and especially Brachet (1960d). It may be mentioned that Coldwater (1930, 1931) determined the content of glutathione in *Planaria agilis* and *Procotyla fluviatilis*, the latter being assumed to have fewer "formative cells" (neoblasts). He found a higher content of glutathione in *Planaria agilis*, which may be significant in the discussion about —SH groups depending upon whether the differences in regeneration rate and amount of neoblasts observed in the two species can be confirmed. He also found that the rate of respiration and the amount of glutathione was lower after X-ray treatment (neoblasts being very sensitive to X-rays, Wolff and Dubois, 1948). Owen, Weiss and Prince (1938, 1939) found that glutathione stimulated regeneration and fission, but that cysteine alone had no influence on the former.

Many other chemicals and extracts of organs have been tried for their effect on regeneration which, with other items, will be discussed later in this chapter.

The early work of Flickinger was dealt with in Chapter 6 on polarity. Coward and Flickinger (1965) studied regeneration with labelled amino-acids in *Dugesia dorotocephala*. In such studies I think the use of species with a pronounced ability to propagate by fission is liable to error, because the fission region may disturb, or at least blur, the pictures obtained with biochemical and "biological" methods; species such as *D. lugubris*, which have a regular time-graded field tapering antero–caudally, do not propagate by fission, and would, I am sure, give more reliable results. I would also recommend for biochemical studies on Flickinger's lines species such as *Dendrocoelum lacteum* and *Bdellocephala punctata* which do not regenerate heads behind the pharynx.

Coward and Flickinger studied the incorporation of leucine-1-^{14}C and ^{14}C-labelled algal protein hydrolysate into protein, of uracil-2-^{14}C into RNA, and the uptake of ^{3}H-thymidine in DNA synthesis; these procedures were done both with intact animals and worms in regeneration. Figure 155, I shows that in intact animals the incorporation of amino-acids is high in the head regions, falls steeply in the pharyngeal region, but rises again in the hind part of the animal, which, let it be emphasized, contains the fission zone. One must ask if the same caudal peak would be found in, say, *Dugesia*

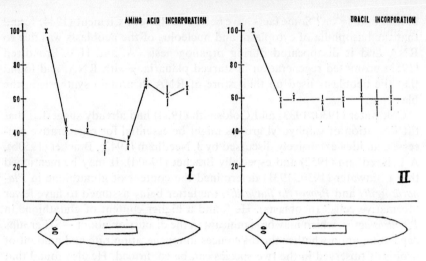

FIG. 155. *Dugesia dorotocephala*. Explanation in text. (From Coward and Flickinger, 1965.)

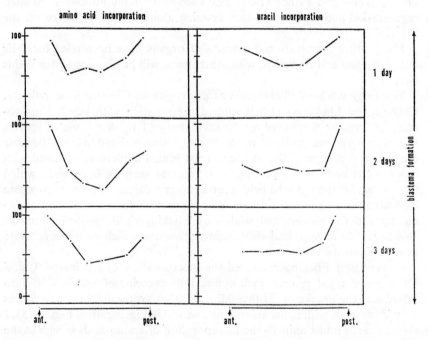

FIG. 156. *Dugesia gonocephala*. Explanation in text. (From Coward and Flickinger, 1965.)

lugubris? The incorporation of uracil is also maximal in the head region, but unlike with the amino-acids there is no second peak (Fig. 155, II). This result is certainly curious. One explanation might be that the RNA in the intact animal had already produced the proteins necessary for subsequent fission processes. Such a hypothesis would explain the findings of Coward

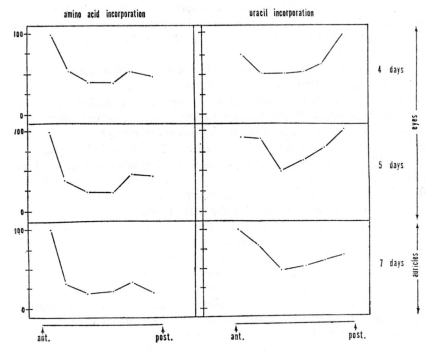

Fig. 157. *Dugesia gonocephala*. Explanation in text. (From Coward and Flickinger, 1965.)

and Flickinger that during regeneration RNA was found to have its peak caudally (Fig. 156) declining almost to normal on the seventh day of regeneration (Fig. 157). It must be admitted, however, that the delay of protein synthesis seems to come considerably after the RNA synthesis so perhaps the time factor is decisive here.

Concerning the experiments on DNA synthesis with ³H-thymidine the authors wrote:

> They reveal a cephalocaudal gradient of DNA synthesis. During the first 2–2½ days the regenerating tail-forming region shows a greater relative increase of ³H-thymidine incorporation than the head, but the anterior tissues always have a higher absolute incorporation of ³H-thymidine. Such a dominance would be expected to exist according to Child's (1941) theory of axial metabolic gradients.

Coward, Flickinger and Garen (1964) showed that stimulation of protein synthesis by nutrient supplementation accelerated both the rate and extent of regeneration in starved planarians, confirming our findings (1953) that RNA accelerates regeneration in starved planarians.

In my opinion it is premature to say anything definite as to the basic mechanism of regeneration on the morphological plane (polarity) from Flickinger and co-workers' ingenious and important analyses. These authors are themselves cautious in their interpretations; they make so many suggestions of value that I strongly recommend a perusal of their paper in the original.

In an important paper Flickinger and co-workers (1963a) endeavoured to extend the work of 1959 on the control of cellular differentiation by regulating protein synthesis. He first examined the incorporation of protein precursors in *Dugesia tigrina* and *D. dorotocephala* to see if Child's gradient theory could be transformed, so to speak, from rate of oxygen consumption to a gradient of protein synthesis. Flickinger incubated 100–250 worms in tap water containing 15 μc $^{14}CO_2$ or 55 μc ^{14}C-glycine for 1–2 hours, and then cut them into three to six equal pieces. Each region was pooled and homogenized. Both procedures indicated maximal incorporation of protein in the head region, so Flickinger deduced that the axial protein metabolism followed Child's gradient. He demonstrated that chloramphenicol, a powerful inhibitor of protein synthesis, abolished the protein gradient, and strengthened his argumentation by removing the mid region of the worms, and, dipping the anterior end in 2% agar containing chloramphenicol or colcemide, he succeeded in reversing polarity in a good many instances.

It is obvious that Flickinger's experiments justify his hypothesis that the level of protein synthesis is connected in some way with head regeneration and polarity. I should, however, say that such a conclusion may be premature as other factors, e.g. neurohormones, must be taken into account before it can be accepted.

In his paper Flickinger also discussed *gene-actions*, but consideration of this must be postponed to Chapter 22. Flickinger did not find that some cells were more specific than others in their uptake of labelled $^{14}CO_2$, and wrote: "It would appear that the axial incorporation gradient of labelled CO_2 and clycine ... is attributable to the metabolic activity of a variety of cell types...".

In a series of papers Lindh investigated biochemical conditions before, during and after regeneration, mainly in *Dugesia polychroa*. He found (1956) that nitrogen-containing substances, fat, sugar and nucleic acids followed specific gradients along the longitudinal axis. The distribution was stable in the intact animal and stability was re-established after complete regeneration. During regeneration DNA and RNA fluctuated during decomposition and re-synthesis. Wounding released periodic variations which affected more than the blastema, as new parts were actually rejuvenated: the content of nucleic acids and polynucleotides was low. Old animals always had a large reserve of synthesized compounds with large amounts of protamine, but little free amino-acids.

Lindh (1957a) found the cytological evidence in the blastema equivocal. All cells in the regenerating part exhibited intensified metabolic activity: after 36 hours to 3 days there was an increase with diffuse basophilia and increased mitotic activity starting lateral to the two main nerve trunks. Later (1957b) he found the amino-guanine ratio heightened when many mitoses were present and deduced that a low amino-guanine content was a sign of rejuvenation. The decline in RNA he ascribed to ribonuclease. During wound repair he found a disturbance of the ratio DNA:RNA and an accumulation of adenylic acid before and during the differentiation period. Many variations in the free nucleotide concentration occurred during regeneration.

Lindh (1957c) surveyed his earlier papers concerning the nucleic acids and metabolism in regeneration without giving any new facts and without supporting Child's gradient hypothesis.

Yamamoto (1957) gave valuable information on the histochemistry of the tissues of *Dendroceolopsis* sp. He found DNA only in the nuclei and polysaccharides in the basement membrane of the epidermis, the rhabdite-forming glands and the muscles. —SH groups were found in all parts of the tissues and alkaline phosphatases in the pharynx and brain (see also Chapter 18).

In conclusion I feel that the claims of the various authors that their results are in accordance with Child's metabolic gradient theory are not clearly substantiated. On the contrary, I think that the most important evidence brought forward supports Morgan's view better. It is, of course, quite clear that the results reveal a metabolic gradient but to my mind the experimental evidence has a deeper basis and events are in fact tied to some genetic influence which, through the intermediary of coded RNA, determines the synthesis of enzymes and other proteins necessary for building up the missing parts of regenerants. I do not think that a mere accumulation of protein determines polarity, i.e. proteins derived from the animal itself or from the environment; if the latter it is, of course, understandable that a high concentration of protein may accelerate the rate of regeneration. I shall discuss the whole matter in the concluding Chapter 23, in connection with the concept of the time-graded regeneration field.

Meanwhile, let us see which varied biochemical methods have been used to study planarian regeneration. These can be conveniently presented under three headings.

1. *Promoting Substances*

Goetsch (1946a,b, 1950) claimed to have found a new vitamin, T, in several insects, especially in termites, but also in *Penicillium* and other fungi. The influence on *Dugesia gonocephala* was to promote regenerative and restorative processes; in *D. lugubris* the rate of head regeneration was increased by 20%. Fraenkel (1953) found the substance which he called carnitin in a wide range of plants and animals. Bahrs (1931), Bahrs and Wulzen (1932), Wulzen (1916) and Wulzen and Bahrs (1935, 1936) found that some factors in mammalian viscera, especially in the cortex of kidneys from guinea-pigs and rabbits fed on alfalfa, accelerated growth in planarians. It might be interesting to see if regeneration also was enhanced. Marino (1957) found placenta brei accelerated regeneration considerably.

Such more or less chance findings, pursued biochemically, might lead to a deeper insight into the physiological chain of events during regeneration. Weimer, Phillips and Andersson (1938) found thyroxine to accelerate eye formation in all regions of *Planaria gracilis*. Owen, Weiss and Prince (1938, 1939) claimed that dibenzathracene and glutathione promoted regeneration slightly in *Dugesia dorotocephala*, but not glutamic acid, glycine or cysteine. It is interesting that Buchanan (1938) claimed that extracts of planarian heads *(D. dorotocephala)* stimulated head regeneration, as did exudates of heads if used within 24 hours. This finding should be considered in conjunction with the induction phenomena discussed in Chapter 9. It has been mentioned before that riboflavine accelerates regeneration in starving planarians. Kanatani (1959a) found that oestradiol in low concentrations (0·01 µg/ml to 0·1 µg/ml) promoted regeneration in *D. gonocephala*, whereas higher concentrations inhibited it. Kanatani (1960b) found that colchicine and demecolchicine promoted fission in *D. gonocephala*. It is uncertain if these findings have something to do with real regenerative processes also.

A promising line of investigation was initiated by Mengebier and Jenkins (1964), who determined the succinoxidase activity in homogenates of *Dugesia dorotocephala*, and found greater activity in fed animals than in those starved for 48 hours. Although without a direct bearing on our present problem it seems that the determination of important enzymes in homogenates of planarians in various stages of regeneration might well provide valuable information.

Coward, Flickinger and Garen (1964) starved *Dugesia dorotocephala* for 5–30 days, cut them into three to six pieces. Regeneration occurred in a solution of 0·005–0·01% yeast sodium ribonucleate, 5% horse serum, 1% amino-acid and 1% chick embryo extract. This medium promoted the regeneration of eye-spots, particularly in the posterior pieces, in which they appeared 1–2 days sooner than in tap water controls. $^{14}CO_2$ incorporation demonstrated increased synthetic activity in posterior levels. Again I must warn against too much stress on this phenomenon, as this species propagates by fission, and therefore contains plenty of neoblasts in the posterior region, thus being presumably more responsive to stimulation. Perhaps the authors had this in mind when they wrote: "... head formation occurred at the cut end of the worm with the highest level of activity of the protein-synthesizing mechanisms". And later: "... the inhibitory role of the head in preventing posterior cephalic differentiation may be due to its capacity to drain substances and co-factors from the more posterior levels."

2. *Inhibitory Substances*

It is, of course, to be expected that many substances, other than poisons, will inhibit or retard regeneration and thus throw light on the biochemistry of regeneration.

Goldsmith and Black (1948) stated that podophyllin in a dilution of 10^{-6} retarded regeneration. McWhinnie (1955) studied the influence of colchicine on regeneration in *Dugesia dorotocephala* and found that weak concentrations inhibited regeneration and regulation, but that this influence was reversible; strong concentrations inhibited the formation of blastemata but not the regeneration of eyes; the medio–lateral gradient was also reduced. She found that a concentration of M/5000 inhibited more strongly when applied on the fourth than on the third or fifth day after transection, which may have something to do with the timing of mitoses. She also found that heteromorphic animals with three to four heads were produced by colchicine and concluded that the effect did not affect differentiation, e.g. she found: "... that the two lateral halves of a piece underwent independent differentiation". Kishida (1961) found that dithiocarbamide inhibited pigmentation of planarian eyes more strongly than thiocarbamide. Kido and Kishida (1960) found the same lack of pigmentation of the eyes with thiourea and iodine, ascribing it to their oxidative effect. Teshirogi and Maida (1956) treated planarians cut transversely into five segments with 0·1 % thiourea and found that both body and eye pigmentation decreased affecting first the regions with high head frequency. Kanatani (1962) studied the effect of nicotine amide and found that a concentration of 10^{-3} M inibited regeneration by reducing the number of mitoses. It was interesting that 5×10^{-2} M glucose or sucrose reduced the retardation of head regeneration caused by demecolcine in *D.gonocephala* (Kanatani, 1959b). He found (1958c) that demecolcine retarded head formation after decapitation of *D.gonocephala* more markedly than colchicine, but that the difference in effect was not perceptible if the chemicals were applied before decapitation, suggesting that in the latter case the chemicals had not penetrated the epidermis.

Studies by Kanatani (1957) showed that crowding retarded fission and produced extra eyes in *D.gonocephala* and also that ammonium chloride produced the same features; the authors deduced tentatively that there must be a close connection between these phenomena, see also p.165.

Goldsmith (1946) transected planarians immediately anterior to the pharynx and found that thiourea retarded head regeneration in concentrations of 0·3–0·5% whereas 1% entirely stopped regeneration, a feature which was, however, reversible, because regeneration was completed when the animals were placed in clean water.

Jenkins (1959) used the goitrogenic substances, thiourea, phenylthiourea and thiouracil. The first and third retarded regeneration slightly, the second effectively. High concentrations inhibited normal development of the sensory lobels and proboscis. Jenkins (1961a), using the Warburg respirometer with *D. dorotocephala*, found that thiourea and thiouracil had no influence on respiration, whereas phenylthiourea had an initial depressant effect, but during regeneration the respiration rate returned to normal and then increased significantly. She discussed possible biochemical pathways, e.g. that the initial depressant action of phenylthiourea might be ascribed to its ability to inhibit tyrosinase. Jenkins further studied (1961) the effect of thyroxine and tri-iodothyroxine and di-iodothyroxine on the regeneration rate of *D. dorotocephala*, but found no significant effect.

Gabriel (1963, 1965), using *Dugesia gonocephala*, found that mercaptoethanol at concentrations above $M/750$ were toxic. In concentrations of $M/1000$ differentiation during regeneration was blocked but the neoblasts migrated. Actinomycin B also inhibited regeneration when administered during the first 24 hours after the wounding apparently due to the influence of the neoblasts, their nucleoli disappearing after the third day of administration. The author thought that the inhibitory action of actinomycin was on the RNA synthetic mechanism.

Kanatani (1959a) found that oestradiol, in a concentration of 0·01 to 0·1 µg/ml, enhanced regeneration in *D. gonocephala*, whereas a concentration of 10 µg/ml inhibited it. Glucose had hardly any effect on this.

Finally, it should be mentioned that Kido and Kishida (1960) found that thiourea and iodine caused a depigmentation of the eyes of planarians, and thought that the iodine oxidized thiourea. Kishida (1961) found depigmentation using dithiocarbamide and thiocarbamide, the former having the stronger effect. Perhaps these findings, on elaboration, might give information as to other biochemical events in planarian regeneration. Smith and Hammen (1963) used *p*-chlormercuribenzoic acid and avidine in planarian regeneration and found only that defective eye-spots developed. The authors thought that these inhibitors of carboxylation were effective in producing defects in planarian regeneration.

3. *Teratomorphs*

It is not surprising that some of the inhibitory substances also produce defects in regenerants; examples have already been given in (2) above. Goldsmith (1946), besides finding inhibition by thiourea, also found a high incidence of atypical regeneration. Goldsmith and Black (1948) found a retarding effect using podophyllin (10^{-6}) with the formation of atypical heads. McWhinnie, using colchicine on *Dugesia dorotocephala*, sometimes found heteromorphoses with three to four heads, and thought it probable that the arrest of mitoses, besides inhibiting blastema formation, brought about irregularities in the further replacement of neoblasts, so establishing more "high-points" (Chapter 5). I think that the phenomenon deserves further investigation, because it might cast some light on the problems of induction.

Kanatani (1958b), using *D. gonocephala*, applied demecolcine, a more powerful inhibitor of mitosis than colchicine, but not as poisonous, in concentrations of M/20,000–40,000. He obtained bipolar heteromorphoses even in pieces of considerable length, an effect enhanced by raising the temperature (Fig. 43). He suggested that demecolcine might influence the nervous system also. Kanatani (1960a) pursued his (1955b) experiments, again using *D. gonocephala*, and found that the forepart of decapitated animals, when divided into four parts and put in demecolcine for 48 hours, developed bipolar heteromorphoses; if held anaerobically for 6 hours, or in M^{-3} KCN, the number of the heteromorphoses was reduced. Kanatani suggested: "... that the rate of respiratory activity associated with phosphorylation constitutes a limiting factor for the formation of bipolar heads ... the reduced respiratory activity apparently leading to the production of tail rather than head at the posterior cut surface". Kanatani (1959) also found that pipolar heteromorphoses produced by demecolcine may be abolished by using glucose and fructose simultaneously with demecolcine. Kanatani found that respiration was increased by using demecolcine, but that this did not occur if glucose was given simultaneously. He suggested that there may be some connection between the increase in respiration and the formation of bipolar heteromorphoses. The phenomenon of heteromorphoses, with the only outward sign being supplementary eyes, was a common feature. Brøndsted (1942) found that LiCl produced several animals with distorted eye formation; this was confirmed by Kanatani (1958a, b). He found that in *D. gonocephala* that LiCl produced supplementary eyes even 16 days after decapitation, which he ascribed to a prolonged lability of the determination situation in the planarian body though it may here be noted that old "normal" specimens often have supernumerary eyes. Kanatani also showed that ammonia, produced by crowding the animals, may result in supernumerary eyes. Kanatani (1959d) analysed the influence of LiCl more closely and found that neither KCl nor NaCl had any effect in producing supernumerary eyes, but that KCl had an augmenting influence on the effect of LiCl in producing abnormalities; NaCl, on the other hand, in high concentrations inhibited the effect of LiCl, and also abolished the effect of KCl on LiCl.

Teshirogi (1955b), using LiCl, stated that the regeneration rate may be inhibited. He also found biaxial heads, again in comparatively long pieces, but most frequently from midpieces of the worms. He (1956a) found that sodium thiocyanate (0·01 %) inhibited both head and tail regeneration more strongly than LiCl in *Bdellocephala brunnea*. A most interesting fact was that in a single specimen, a postpharyngeal piece regenerated a head after the application of NaSCN. It is very hard to interpret this finding, and it should certainly be reinvestigated because, if confirmed, it might cast light on the well-recognized phenomenon that *Bdellocephala* is unable to regenerate a head from this part of the body.

In conclusion it may be said that our knowledge of how chemicals produce heteromorphs is very scanty, and in the few cases we know biochemical interpretation is obscure. During regeneration the substances employed may influence many factors, e.g. respiration, mitosis, migration of neoblasts, neurohormones, acceleration or retardation of enzymatic processes and so on. Therefore the normal interaction between these and other yet unknown factors may certainly be disturbed resulting in heteromorphoses. A long time ago Runnström used the term "Gefälleanarchie" to describe this in a general way.

The investigations discussed under heading 3 in this chapter, together with the phenomena discussed in Chapter 14, are, I think, very promising for deeper biochemical analyses.

In view of the possibility that the presence of various micro-organisms in and on the planarian body might interfere with biochemical determinations, the work of Miller, Johnson and Millis (1955) and Miller and Johnson (1959) is of interest; these authors attempted to raise planarians in germ-free cultures.

CHAPTER 20

BIOPHYSICS AND PLANARIAN REGENERATION

THIS chapter covers some of the biophysical influences under three headings: (1) Irradiation. (2) Electrical stimulation. (3) Magnetism.

1. *Irradiation*

Bardeen and Baetjer (1904) used X-irradiation (without stating doses) on *Dugesia tigrina (maculata)* and *D. lugubris*. After transection, regeneration of a head or pharynx occurred depending on the level of the section. Transected animals died 20–22 days after irradiation. Intact animals died after 1 month without exhibiting external signs of abnormality, with necrosis starting at the head and going caudal. If irradiated animals were transected shortly before dying no regeneration followed. The authors stated that X-irradiation had a powerful inhibitory effect upon cell-reproduction in planarians.

Shaper (1904) used X-irradiation and radium on both embryonic and regenerative processes with results the same as Bardeen and Baetjer. Hinrichs (1926) found Child's susceptibility gradient using X-irradiation and ultraviolet. Gianferrari (1929), also using X-rays, found the same disintegration gradient in *D. polychroa*, but not in *Polycelis nigra*, a difference worth noting and reinvestigating.

Strandskov (1934, 1937), in *Dugesia dorotocephala*, using doses from 172 to 864 r, found that the susceptibility gradient coincided with the "chemical" one (Child, etc.). Surviving animals in most instances regenerated some of the lost tissues, but the regenerating faculty was diminished on increasing the dose. Van Cleave (1934) used X-irradiation on *Stenostomum tenuicauda* and found that the animals survived a dose of 30,000 r when exposed over a period of 9 hours, but succumbed to 20,000 r given in 36 minutes. He found that the surviving animals could regenerate new heads and reconstitute themselves entirely before cytolysis. These results should be reinvestigated because they are the opposite of those of Dubois; if confirmed there must be a striking difference in resistance to X-rays in *Stenostomum* and paludine triclads.

It was previously stated (Chapter 7) that Curtis was aware of the significance of "formative cells" (neoblasts) for regeneration. Curtis and Hickman (1926) treated planarians with X-rays and radium and suggested that destruction of the formative cells by irradiation was the cause of non-regeneration. They believed that there was complete destruction of these cells, regarded as embryonic and therefore more sensitive to irradiation than other cells, e.g. in *Dugesia agilis* no regeneration followed and the animals died 4 weeks after irradiation. Curtis (1928, 1930) confirmed this.

Schewtschenko (1938), using 2500–3500 r, was the first to use partial irradiation. He found that non-irradiated segments regenerated, whereas irradiated ones cytolysed. Weigand (1930), in *Polycelis nigra*, found that with radium 5–15 minutes' irradiation delayed regeneration of eyes, namely 12–14 days against 8–10 days of the controls. Stronger doses caused deformities and diminished blastema formation. All the animals in these series were transected immediately after irradiation. If the transection took place 1 day after 5 minutes' irradiation, regeneration was completed, although eyes appeared 4–5 days later than in controls; if transected later no eyes were found. *Dugesia torva* and *D. lugubris* showed the same phenomenon, although *D. lugubris* was more resistant to irradiation. If the animals were transected and irradiated $1\frac{1}{2}$–9 days later, the longer the interval between transection and irradiation the better the eye-formation. Weigand believed from this that the blastema ("Regenerationskegel") developed resistance with increasing time. Using a 5·5-mm aluminium filter, which only allowed hard gamma rays to penetrate, only a few deformations occurred. Weigand did not find mitoses in the irradiated animals, but sometimes "zerbröckelte" chromatin, whereas in non-irradiated animals many mitoses were seen (see Chapter 7). Weigand used this phenomenon to discuss de-differentiation.

Weserve and Kennedy (1935) X-irradiated 25–50 *Dugesia dorotocephala* starting simultaneously in each experiment but varying the dose from 4 to 12 HED by 95–100 kilovolts without a filter; 23 hours after irradiation both the fore- and hind part were removed. After 3 weeks heads were regenerated, but they broke away from the regenerate. After 34 days the dissolution of the animals was proportional to the duration of the irradiation. Pigment never formed in the regenerated heads. After 34–38 days all animals with 8–12 HED died.

In Chapter 7 the classical paper of Dubois (1949c) and others of the French school were discussed. Dubois used a Coolidge tube with a tungsten anti-cathode measuring $1·9 \times 1·9$ mm without filter at 60 kilovolts and 4 milli-amps; the distance from the anticathode to the upper surface of the animals was 8·2 mm; the dose per minute was 320 r. Irradiation lasting 25 minutes, with a total dose of 8000 r completely inhibited regeneration and caused death in 3–7 weeks. Doses of 2500–5000 r were in the lower limits for inhibition of regeneration. The animals were more resistant in spring and summer— a phenomenon well worth investigating. Ortner and Seilern-Aspang (1962) also used 8000 r in their experiments.

Kratochwill (1962) X-irradiated *Dugesia gonocephala* at 170 kilovolts and 19 milliamps, 400 r per minute with a total dosage of 8000 r. No regeneration followed but interestingly this did not inhibit the migration of neoblasts to the wound; only their differentiation was inhibited, and then only if decapitation followed irradiation immediately. If, however, irradiation was given 24 hours after decapitation, normal head regeneration followed. Kratoch-

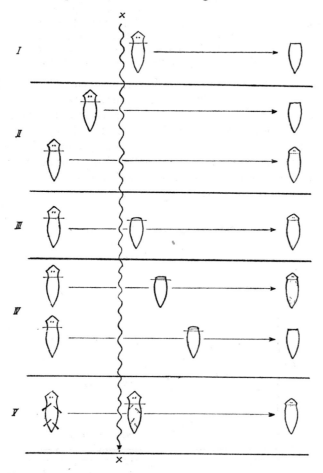

FIG. 158. *Dugesia gonocephala*. Explanation in text. (From Kratochwill, 1962.)

will deduced that during these 24 hours: "... ein Vorgang abgelaufen sein muß, der aus den Regenerationszellen im Blastem ermöglicht, interzelluläre Wechselwirkungen aufzunehmen und gemeinsame Differenzierungen zu bilden". This "Differenzierungsbereitschaft" was, however, a phenomenon which also occurred even if wounding was carried out without decapitation. The "Differenzierungsbereitschaft" involved all the neoblasts in the body. These findings of Kratochwill were very interesting because they opened up new aspects of the biochemistry of neoblasts when animals were stimulated by injury. But:

> Nach einer bestimmten Zeit, gleichzeitig mit dem Auftreten von ersten Differenzierungen im Blastem, klingt diese Differenzierung der im übrigen Körper liegenden Regenerationszellen ab. Dies zeigt die negativen Resultate der Re-regenerationsversuche, bei denen die zweite Amputation später als 57 Stunden nach der ersten durchgeführt wurde. Es muß unentschieden bleiben, ob die Differenzierung der im Körper gelegenen Regenerationszellen nach einer bestimmten Zeit von selbst abklingt, oder ob sie von den im Blastem auftretenden Differenzierungen unterdrückt wird.

Figure 158 shows schematically the various experiments which led Kratochwill to his deductions. Personally, I think that his latter alternative is the right one because it agrees with the inhibition discussed in Chapter 10.

Kratochwill hypothesized on the influence of ionizing irradiation, and suggested that irradiation was perhaps followed by a migration of calcium ions from the surface of the neoblasts, so diminishing the adhesion between these cells and inhibiting the co-operation which otherwise should have led to the building up of an organized blastema.

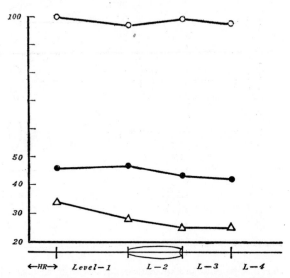

FIG. 159. *Dugesia gonocephala*. Head-frequency of pieces of X-irradiated worms 20th day after cutting. White circles: control. Solid circles: 1200 r, open taringles: 1500 r irradiated pieces. Ordinate head-frequency index, abscissa body level. (After Teshirogi, 1961.)

Teshirogi (1961), in *Dugesia gonocephala*, found that irradiation with 1000–1500 r reduced the regeneration of heads by 30% or more (Fig. 159). It is worth noting that the head-frequency curve in Teshirogi's experiments was much more flattened than generally seen in the literature.

Teshirogi transplanted rectangular ganglionic grafts of non-irradiated animals into irradiated animals posterior to the pharynx, with induction of new pharynges both anterior and posterior to the graft, irrespective of its orientation. If the transplanted grafts were triangular, new pharynges were induced only from the bases of the grafts, again without relation to their polarity. At the same time a head and a post-cephalic outgrowth developed both from rectangular and triangular grafts.

It seems that two major deductions may be made from these experiments: (1) that non-irradiated cephalic pieces induce pharynges in irradiated tissues, and (2) that polarity is not retained in the irradiated tissues.

Teshirogi (1963a) continued his irradiation experiments in *Bdellocephala brunnea*, a species which, like *B.punctata*, is unable to regenerate a head posterior to the anterior part of the pharynx. Teshirogi found the same as Stéphan-Dubois (1961) in *Dendrocoelum lacteum* in that neoblasts migrated from the posterior parts of the body (unable to regenerate head) through an irradiated segment transplanted to the anterior part, and here regenerated a head provided the transplant did not contract. Teshirogi also found that head regeneration was very delayed, as was to be expected. No mitoses were found in the irradiated parts. In non-irradiated grafts the number of neoblasts was diminished during head regeneration, whereas the number of neoblasts in the posterior part remained unaltered. In contrast to this, the number of neoblasts was diminished in the posterior part when head regeneration took place from the irradiated graft. Teshirogi was right in saying: "It is noteworthy that the X-irradiated anterior component when grafted on a non-irradiated piece did not get disintegrated during the regeneration." Could this mean that the neoblasts in some way released substances which restored the viability of irradiated tissues? It seems to me that this phenomenon deserves further investigation because it might prove to be of very general biological importance.

Ionizing irradiation from the thermal water of Gastein inhibited regeneration in *Dugesia gonocephala* considerably (Haslauer, 1962).

It is of paramount interest to determine the threshold dose of irradiation causing damage in the various cell types in the planarian body and also the threshold of damaging irradiation to cells in their various physiological states. There is much to be learnt. Some indications have been hinted at in Chapter 7. The problems involved have been briefly investigated by Fedecka-Bruner (1961, 1965).

It may be mentioned here that Hinrichs (1924b) found that the planarian body disintegrated according to the axis of metabolic activity (Child) when ultraviolet was used in *Dugesia dorotocephala*.

In contrast to Hinrichs, Merker and Gilbert (1932) found that cytolysis often began at places other than the head under the influence of ultraviolet. Using *Dugesia dorotocephala* and *Dendrocoelum lacteum* the authors ascertained that cytolysis began at places where the ultraviolet struck the animal perpendicularly and where cuts or needle-pricks had been inflicted. The authors further stated that resistance to ultraviolet decreased with increasing duration in captivity.

2. *Electrical Stimulation*

It is a reasonable inference that a voltage gradient should exist in the planarian body, falling off caudally, if a metabolic gradient, exemplified by that of oxygenation, were present, as maintained by Child. Several authors have tried to find an electrical polarity conforming with the morphological one in planarians.

Such gradients have been found in many instances in the living world (Becker, 1961) but the records concerning planarians are conflicting.

Lund (1921, 1925, and co-workers, 1949) believed in a close correlation between electrical and morphological phenomena, and were evidently impressed by Child's metabolic gradient hypothesis in planarians.

It is commonplace that cell metabolism is linked to electric events, but this is certainly not enough to explain morphological polarity in planarians. Diversity of cell metabolism may very well be concomitant with a corresponding diversity of electrical potentialities, but need not necessarily be so. The planarian body is a vast complex of various cell types which contain a great variety of intricate metabolic activities, but it seems difficult to see how this should be responsible for an overall antero–caudal gradient. Another thing is that electric currents applied to the animal externally may strongly influence its morphology. For a discussion of this see Robertson (1927) and for a more general survey see Rosene (1947), Evers (1947) and Nakazawa (1961).

Dimmit and Marsh (1952) and Marsh and Beams (1952) closely examined regeneration in *Dugesia tigrina* using electrical stimulation. Heads and tails were removed and discarded; a slit marked the anterior end of the remaining body, which was then imbedded in 2% agar. The experiments showed that the potential fall across the regenerating piece was an independent variable on the effect of the electric field upon determination of the polar axis. In controlling axial polarity in *Dugesia* the applied electric field operated as a morphogenetic field and conformed to all the elements of the morphogenetic field concept (Weiss, 1939). The authors thought that the patterned discharge of self-generated electrical energy through the conducting tissue fluids and cells acted as a naturally controlling complex electrical field in morphogenesis.

In other experiments decapitated animals were cut into 2–5 mm pieces, which, in the authors' opinion, would be too short to regenerate Janusheads. The pieces showed cathodal galvanotropism and galvanotaxis. Although many pieces in agar pointed their anterior end towards the cathode, some underwent a reversal of original polarity at $21 \cdot 62$ mA/mm^2 (average) when the potential gradient in mV/mm was 10·31 current density. The authors further found that no regeneration delay or inhibition was observed other than that implicit in control of the polar axis and that the morphogenetic result of the effect of the current was unrelated to the piece length or the original body position.

These experiments are of interest for general physiological reasons, but as far as I can see they tell us nothing about inherent electrical polarity in the intact and uninfluenced planarian body in relation to morphological polarity. On the other hand, many possibilities as to the reversal of polarity in regenerating planarians under the influence of an applied electrical current present themselves, e.g. the influence on neoblasts, neurosecretion and so on.

Flickinger and Blount (1947), in a paper with a valuable theoretical discussion, could not find in the intact *Dugesia dorotocephala* any electrical polarity conforming with the morphological polar axis, nor any respiratory gradient (see Chapter 17). If an e.m.f. was applied along the longitudinal axis, respiration was raised towards the cathodal pole, and so a real respiration gradient was created.

In conclusion it may be said that differences in electric conditions in the planarian body—however they may be generated by the metabolism of the cells— do not coincide with morphological polarity. So neither these experiments nor those concerning respiration levels support Child's hypothesis.

3. *Magnetism*

As far as I know, no experiments pertaining to a possible influence of magnetism on regeneration have been made. It should, however, be mentioned that Brown (1962a) has shown *Dugesia* to exhibit a response to weak magnetic fields in the range of 0·17–10 gauss. In a study of the difference between the north and south poles the animals were able to resolve intermediate angular orientations of fields with remarkable precision suggesting the perceptive mechanism to be specifically adapted to as weak a field as the geomagnetic one. Brown and Park (1964) studied seasonal variations of gamma-taxis.

CHAPTER 21

ENGRAMS IN PLANARIANS

THE experiments of McConnell and his co-workers, pupils and critics, on the transfer of "memory" in planarians are world-famous.

Although I am not in a position to give an expert opinion on these, I nevertheless feel that a short summary should be given in this book because the experiments involve regenerative processes.

The reader should start with a perusal of the paper by McConnell (1966) as they will find in it, (i) a very critical and thought-provoking discussion of the term "learning"; (ii) a survey of "learning" and "memory" in Protozoans, Coelenterates, Plathyhelminthes, Annelids, Molluscs and Arthropodes; (iii) an extensive and complete bibliography; (iv) a survey of the relevant experiments with planarians which will give those of us who are more or less ignorant in this rather intricate field an excellent introduction.

It is clear that the procedures of conditioning of planarians to various stimuli involve measurement of very complicated physiological parameters. This was found after Johnson's and McConnell's first experiments in the early 1950's, reviewed in McConnell's 1966 paper, pp. 117–21.

Highly relevant to the problems of planarian regeneration is the fact that McConnell, Jacobsen and Kimble (1959) showed that the "engram"—whatever the ultimate definition—(mostly in *Dugesia gonocephala*?) was retained after regeneration in both head and tail parts, a fact corroborated by others. It was concluded that the engram must be stored in some chemical way (i.e. the well-known Hydén hypothesis). Untrained planarians fed on trained ones cut in pieces "learnt" better than controls fed on pieces of untrained ones.

It was deduced that RNA could be a "transfer" agent for "memory". To test this, RNAase was added to the pond water in which conditioned animals were allowed to regenerate. Curiously enough, the posterior halves "forgot" their conditioning, while the anterior ones retained it. In other experiments RNAase was injected into conditioned animals "erasing" "memory". Further, RNA was extracted from conditioned planarians and injected into untrained ones; the result was that these worms learnt better than uninjected ones. If the RNA was treated with RNAase, it had no effect.

McConnell justifiably asked: "Is the 'transfer action' specific or general? That is, are specific 'memories' being passed from one organism to another, or is it merely a generalized 'activity level' that gets transplanted? I have suggested elsewhere that both types of effect might be transferable...".

I, myself, am not prepared to go into a deeper discussion on the tranferability of "memory" on a molecular basis. Besides the very exciting experiments with planarians (which may be followed in the serious yet humorous journal *Worm Runners Digest*, edited by McConnell) the most critical experiments I know are those of Fjerdingstad, Røigaard-Petersen (former pupils and co-workers of mine in other biological fields) and Nissen (1954 and 1966) using rats. It goes without saying that experiments on "molecular memory or engrams" are likely to be one of the most significant biological fields in coming years.

I would like to urge the following experiments in planarian regeneration. A study of the transfer mechanism should be extended to species which never propagate by fission, i.e. *Dugesia lugubris, D.polychroa*, to eliminate the errors latent in species apt to propagate by fission, where collections of neoblasts near the fission zone may give rise to errors.

Experiments should be extended to species such as *Bdellocephala punctata* and *Dendrocoelum lacteum* where head regeneration does not occur behind the pharynx, to determine whether such "restriction" is localized only in the head-producing part or is present elsewhere in the body.

It is obvious that conditioning involves the nervous system, because this transmits external stimuli from the sensory organs. We cannot be sure whether other tissues in planarians, especially the neoblasts, are involved. To determine this, I would recommend the investigation of trained species in which we know the distribution of neoblasts in the intact body (p. 119, Fig. 103). Conditioned worms could be cut into transverse pieces according to the curves of distribution of the neoblasts, and the regenerants used as chopped food for non-conditioned worms. This would show if there was a systematic variation of neoblast distribution coinciding with other developmental criteria (using non-conditioned animals as controls).

It would also be of interest to see if foreign RNA could stimulate "learning" in planarians. It is known that RNA (from yeast) stimulates regeneration. Could planarians regenerating in water containing pure RNA from other organisms reach a given developmental stage faster than controls in pure water?

To my mind if we really wish to be sure that RNA is the "memory" molecule it is absolutely necessary to use pure RNA free from proteins, by up-to-date methods.

In writing this small chapter a remarkable idea occurred to me: the basis of the morphological pattern is intimately bound up with DNA, as are the physiological behaviour patterns because morphology and physiology are genetically complementary.

We know that external factors may transform morphological patterns into new phenotypes, and we have reason (Chapters 10 and 22) to believe that cytoplasmic factors release the time mechanisms of gene actions. Could it be that external stimuli via the sensory organs, nervous system, muscles, etc., could influence first the cytoplasm of neoblasts in the conditioned planarians, thence the time-scheme of the action of the genome in the neoblasts and thereafter the pattern be transferred to the physiological behaviour ("memory") of the conditioned animals?

Mad as this idea might be, it is based on the often neglected principle that we must not separate the morphology of animals from every variety of their physiological expression, i.e. their behaviour in its widest sense. As human beings we are no exception to this basic biological rule. Therefore this remarkable idea of mine should perhaps be paid some attention, even in the interpretation of experimental results in other mammalian species—in rats for instance.

CHAPTER 22

THE GENETIC CODE AND PLANARIAN REGENERATION

IN PLANARIANS we are faced with the intriguing phenomenon that a very tiny fragment of the animal is capable of transforming itself into a complete unit with all the "adult" animal's morphological and physiological characteristics; in fact, every part of the planarian body has the capacity to regenerate a complete specimen typical of its own species.

We are faced here with two major phenomena. The first is the spectacle of regeneration itself, that is the determination and arrangement of the cells in the fragment to form a new whole individuum. The second, less conspicuous, is the nature of the forces present in the intact body inhibiting each part from regenerating, when the animal is complete.

The forces underlying these two phenomena are in a way complementary. We have discussed induction and inhibition earlier, especially in Chapter 10. Here I propose to discuss regeneration only from another angle.

We have good reason to regard neoblasts as totipotent. They must therefore be equipped with the entire genetic code of the species. Hence it is one of our tasks in the study of regeneration to unravel both the machinery which releases the potentialities of the neoblasts, and that which underlies the unfolding of the genetic code of these cells during embryogenesis from the zygote. The first task we have already discussed in Chapters 7 and 9. The second is even more formidable than the first. In the egg the genetic code works —metaphorically speaking—on its own initiative; it starts right from the beginning directing the processes more or less dictatorially and co-ordinating them gradually as they come into being; the genetic code in the egg already has its blueprint, as an architect has his plans fully prepared beforehand enabling him to construct his building without interference from surrounding structures. In regeneration the situation is more complicated as no such blueprint can be ready, as it can only be made after the creation of damage. Moreover, the blueprint must be made according to the site and extent of the damage, in co-operation with existing remnants of the body plus the genetic codes of numerous neoblasts, all of which have the power to regenerate the whole. If the neoblasts did not utilize their own genetic codes, they would have to be marshalled into meaningful co-operation in some other way.

Let us once more compare embryogenesis with regeneration. The neoblasts may be considered as blastomeres in a blastula of a highly organized type. It would open interesting and important perspectives if a technique could be found for growing isolated neoblasts in tissue culture. We should very much like to know if isolated neoblasts could be induced to grow and multiply; and if so, would this result in the formation of a new animal—after the scheme, say, elaborated in *Dictyostelium*? Or would they, as seems to me more likely, need stimulation from substances in the adult body to force them to differentiate. Several of the principal problems in regeneration might be solved if such a technique could be developed.

In the meantime we have to content ourselves by investigating how neoblasts behave during "normal" regeneration. Our scanty knowledge in this sphere has been discussed already in Chapters 7 and 8.

It would be premature to speculate here on the possible ways the genetic code in the neoblasts may act during the various differentiations of these cells. Hypothesis, however, comes before experiment and it may therefore prove to be of value to make some guesses.

In Chapter 5 it was discussed how regeneration of heads from lateral parts was inhibited by the "high-point". From this we can deduce that other factors besides the genetic code determine the actual fate of neoblasts during differentiation. I think it reasonable to deduce that in the head, and all other regeneration elsewhere in the planarian body, a guiding force is set up which regulates the proceedings in an orderly manner. It has been shown by Brøndsted (1946) that such a guiding force was produced by injury giving rise to the formation of a blueprint. A tentative hypothesis as to its function, proposed by Brøndsted (1954), will be discussed in the concluding Chapter 23 in the light of later experiments and hypotheses originating from other work.

We must emphasize that the same neoblasts which, after decapitation of the animal, form a blastema which develops into a new head, would have formed a blastema developing into a tail if they had happened to be located only just anterior to the transverse cut, that is, in the hind part of the anterior piece of the animal instead of foremost in the posterior piece (Fig. 47). From this fact it is evident that some factor in the body influences the neoblasts prior to the unfolding of the code.

We know nothing about the dynamic state or stage of the genome in the neoblasts or their possible functions in the intact body. Here it is appropriate to quote Holzinger (1963): "An 'undifferentiated' or 'de-differentiated' cell is one of whose activities the embryologist is ignorant." From the embryonic appearance of the neoblasts (few, if any, signs of differentiation, plenty of cytoplasmic, dispersed ribosomes, large nuclei and nucleoli) one is tempted to consider the genome of these cells as quiescent, but poised to release the code. Perhaps some blocking factor in the cytoplasm prevents the genome unfolding; possibly a chemical block analogous to that which must be supposed to exist in the ripe egg before fertilization; probably inhibiting forces in the already formed and normally functioning body of the worm maintain equilibrium between all cells—as they needs must do in every fully grown system as most individual animals are constituted. These forces are almost certainly serological in nature, acting preferentially on cell surfaces. When a part of the body is removed, the serological equilibrium must be disturbed, and the inhibiting forces may be rendered ineffective by "wound hormones". One or more substances of this rather vague category might act as triggers, first on the cell surface of the neoblasts, then on their cytoplasm, and then, with the cytoplasm as mediator, on the DNA of the chromosomes.

A first approach to this problem might be to find out if the RNA in the nucleoli and cytoplasm of the neoblasts increases. This seems so in amphibians during gastrulation and neurulation (Tencer, 1961) and there are similar indications in planarians (in Chapter 7). This brings to mind T.H. Morgan's (1904a) words: "The initial differences in protoplasmic regions (of the egg) may be supposed to affect the activity of the genes." A careful electron microscopic investigation of the number and reorientation of ribosomes in the cytoplasm and nucleoli of neoblasts in the first hours and then some days after injury might give some useful information.

An increasing body of facts suggest that substances in the cytoplasm may influence the genome, and a few instances may be given.

Hennen (1963) transplanted embryonic *Rana pipiens* nuclei into the cytoplasm embryos of *R.sylvatica*; after approximately ten to twelve divisions the *R.pipiens* nuclei were retransplanted to *R.pipiens* cytoplasm, and were found to be restricted in their capacity to promote normal development. Investigations showed that the chromosome complement was abnormal, both as to numbers and the form of the single chromosome.

Markert and Ursprung (1963) injected small amounts of protein fractions from the livers of adult *Rana pipiens* into zygotes of the same species. Albumin fractions and to a lesser degree histones stopped development at late blastula stage. Nuclei from these blastulae transplanted through seven generations showed no recovery. The authors thought that the effect was primarily on the chromosomes and was replicated during cell divisions. Both normal and abnormal chromosome complements could be seen in the same individual. They concluded that this demonstrated that a replicated persistent change in the nuclei had been produced by the protein injections, and that the injected proteins had produced a change in chromosomal DNA akin to mutations or entered into inhibitory combination with DNA and been replicated along with the DNA throughout succeeding mitoses.

Niu (1963) wrote: "Liver RNA was capable of inducing (mouse) ascitic cells to manufacture tryptophan pyrrolase (TPO) and glucose-6-phosphatase but kidney RNA was unable to do so. On the other hand, kidney RNA was found to induce the biosynthesis of a greater amount of 1-amino-acid oxidase than 1-RNA...". The relative stability of this property suggested that the

RNA-induced change occurred at the gene rather than at the cytoplasmic level. The author thought that this disclosed an affinity of RNA for chromosomes, which were thereby activated, producing a general increase of newly synthesized RNA in the DNA-dependent cytoplasm.

Stich (1959), working with *Chironomus* and *Acetabularia*, said: "It may be that cytoplasmic factors regulate the activity of different chromosomal loci, resulting in the production of nuclear substances with different composition."

Yamada and Karasaki (1963) demonstrated the transformation of iris cells to lens cells during lens regeneration in the newt; ^{3}H-cytidine and ^{3}H-uridine were incorporated into nuclear RNA during the first few days after lens removal. They wrote: "This may mean that whatever the nature of the factor which affects the iris cells, it quickly affects the synthetic activity of the nucleus. It is probable that the RNA synthesized in the nucleus of iris cells is associated with the general events of nuclear replication as well as with production of the cytoplasmic apparatus for protein synthesis."

In view of this and other evidence it may be that the genome of the neoblasts in the blastema is somehow influenced by substances emanating from the old tissues.

Owing to the extreme difficulties in isolating sufficient neoblasts it is hardly possible to investigate the chemical nature of their RNA. Owing to the despiralized state of chromosomes in resting neoblasts it also seems impossible to get information about alterations of the chromosomes—e.g. "puffs" or "loops"—which might prove to be of great value as to the problem of possible influence on the genome during the determinative stages of the neoblasts. For the time being we must content ourselves with using indirect methods. Perhaps the technique of Melander (1963) might be of some use in this matter. See also the very interesting paper of Benazzi (1966).

Investigations by autoradiography with labelled amino-acids trapped by the neoblasts should be of great value. This technique might show a difference in the incorporation of amino-acids in the various sites in the blastema, in relation to (i) the distance from the "high-point", and (ii) the forecasting of the beginning of differentiation of neoblasts into specific cell types.

By feeding planarians with labelled amino-acids—preferably those with —SH groups—it might be possible to see if the amino-acids are taken up specifically by the cytoplasm of neoblasts forming the blastema, subsequently penetrating the nuclear membrane.

Even if information was obtained along these lines, we have a long way to go before we can understand the interaction between the possibility of activated genome in the neoblasts, epigenetic biochemical stimuli from the old tissues and organs, and mutual interactions among neoblasts in the blastema.

I should like to quote Holtzer (1963): "Embryological thinking is largely sociological, dealing as it does with mixed cell populations. To date the experiments which constitute the core of embryological thinking cannot be translated or interpreted in biochemical terms." This statement has an even greater significance in regeneration, where the old tissues are involved providing extra biochemical X's, so confusing the picture even more than in embryological processes.

An alternative possible interpretation of the differentiation process is, of course, that the genome in the neoblasts is not involved. It might be postulated that epigenetic biochemical interactions between old tissues and the cytoplasm of the cells in the blastema would provide a sufficient basis for regenerative differentiation. If we follow this train of thought we must accept the fact that we abandon some basic ideas, such as the necessity of the genome for the formation of enzymes in the cytoplasm and the biochemical chain: gene (DNA), messenger RNA, ribosomal and transfer RNA proteins. For the time being the evidence is very much in favour of the reality of such a chain; it is therefore best to accept the idea of co-operation of the genome in differentiation processes in neoblasts, but also of co-operation of inhibitory forces in the developing blastema.

In Chapter 10 we have clear-cut evidence of the existence of inhibitory forces but how should we picture the integration of genome and inhibitory forces in the blastema?

It is necessary to emphasize that differentiation of neoblasts in the blastema is a two-fold process: (i) the single cell is relegated to some specific task, e.g. nerve, eye or muscle cell, etc.; and (ii) all single cells are differentiated according to a *pattern*. We can therefore conclude that in normal regeneration these two processes are complementary. This is very important as we are too apt to concentrate either on cell differentiation, or on the pattern. In all normal morphogenesis, as opposed to in malignant growth, the two are, of course, inseparable.

In the blastema we have an assemblage of totipotent neoblasts, which, during the first 2 days, are transformed and integrated into the remaining organ system forming a functional animal. Depending upon the level of the incision organs have to be regenerated *de novo*, and this requires a blueprint for its development. It is our task to find out how this blueprint is organized and how it works.

CHAPTER 23

ATTEMPT AT A SYNTHESIS

OUR concluding task is to try and integrate the facts set out in the preceding chapters to make a coherent picture, and to see if the many often quite contrary interpretations deduced from them might be reconciled into a comprehensive working hypothesis. I am vividly aware that such a task must of necessity be approached impartially and though liable to severe criticism always must be based strictly on the known experimental facts.

1. *In planarians, as in most animal phyla, the greatest evolutionary step was the development of bilateral symmetry, giving the basic morphological features found in the genome.*

All experiments dealing with regenerative problems in planarians are somehow or other influenced by this basic hereditary trait and in every circumstance the regenerant strives to attain bilateral equilibrium. This inborn urge is even stronger than that inherited for the establishment of antero–posterior polarity, a drive which may be greatly influenced epigenetically, e.g. by chemical stimuli (Chapter 6). As far as I know, no experiments have been done which illustrate suppression of the drive to establish bilateral symmetry. In this book are presented numerous instances, to the contrary of median splitting of the planarian body in foreparts and tails, of "arms" and of heteromorphoses by T-cuts, by longitudinal and transverse transplants, all always resulting in bilateral structures (Fig. 33).

At first glance there seems to be a strong contradiction between the well-established fact that chemical forces emanating from regenerating organs inhibit regeneration of their like (Chapter 10), and the fact that the organs of a medially split animal do not inhibit regeneration of the same symmetrical organs.

How is the apparent contradiction in regenerative processes to be understood?

Each half of every bilateral animal is in a way a distinct individual but requires close apposition to its complementary half to function properly. The genome in the egg segregates the cytoplasm into left and right parts, which must be the expression of a slight biochemical difference. Cutting the animal medially shows that all symmetrical organs regenerate without inhibition from their symmetrical opposites. Therefore organ-inhibiting substances only inhibit organs in the lateral half alone of the left or right half-animal.

Only if we adopt the hypothesis of a genetically determined left and right animal can we understand the non-inhibition of symmetrical structures until after regeneration when inhibition of left by right and vice versa is restored.

We are confronted here with the same phenemenon as in a well-known experiment with amphibian eggs: if the first cleavage goes through the grey crescent we may separate the two halves, and both will regenerate the missing symmetrical half. If, instead of separating, we kill the nucleus of the right or left blastomere, the other blastomere will only regenerate a half-animal, suggesting that the cytoplasm of the blastomere with the dead nucleus, remaining in close contact with the uninjured one, contains inhibitory substances.

In planarians time must be considered in symmetrical regeneration. On p. 104, Fig. 86 some experiments quoted showed that when a continuous blastema was formed only one head was regenerated. The experiment on p. 40, Figs. 34, 35 showed that where time was involved the regeneration rate (measured by the appearance of eyes) declined caudally and laterally in both half-animals. Each half of an animal split medially will regenerate its mirror half with the same time-graded field as its director of regeneration. If the transplants are slid lengthwise in two half-animals separated by a median longitudinal cut, then the degree of transposition of the two symmetrical time-graded fields will influence the result in respect of the head or heads of the homochimera. Each half-animal will start to regenerate its own half-head, but the result will depend on the speed of regeneration in the opposite half. If the shift is slight, regeneration of the symmetrical eye from the first half will be inhibited by the eye already regenerated in the second, because normal equilibrium between the two half-animals is attained before a symmetrical eye in the first has had time to regenerate. With transposition, i.e. of the two time-graded fields, inhibition from the "weaker" field in the second half-animal may be so weak that the half decapitated in a place of "strong" time-graded field might have sufficient time to regenerate its own symmetrical eye. If the longitudinal shift is considerable, it will suppress the regeneration of eyes in the half with the "weak" field.

What is the influence of this hypothesis of genetically determined symmetrical half-animals upon our understanding of the various regeneration phenomena?

In the first place, neoblasts must contain the entire genome because even if located exclusively in the one half, they are able to regenerate the opposite half. Secondly, they can segregate themselves in one half-blastema by a left and right direction of their migratory potencies. How this is done is at present wholly unknown. Thirdly, inducing and inhibitory substances are generated in the blastema (Chapters 9 and 10), but only if the blastema is formed.

from a transverse incision through the time-graded field in both halves at an almost identical level. Fourthly, inhibitory substances appearing during regeneration act only in one half and not in the other. The experimental facts thus show that the symmetrical halves of bilateral planarians are in reality two animals (left and right), indicating why organs *during regeneration* inhibit their like in the one half but not in the other.

This conception is strengthened by some as yet unpublished results of heterotransplants between *Dugesia polychroa* and *D. lugubris*. Figure 23 shows such transplants which, when decapitated soon after transplantation, regenerated a common normal head consisting on the one side of *D. polychroa* and on the other of *D. lugubris*. Figure 24 shows that the latter eventually regressed to a thin lateral rim attached to the head of the *D. polychroa* section.

These experiments show that incompatible serological factors occur but only after some time after the transplantation. The neoblasts of the two half-animals appear indifferent towards one another, like the early embryonic cells of, say, *Triturus* and *Rana*.

I should mention that I have never succeeded in obtaining successful results from the following type of experiment: two animals were separated into four by median cuts; the left half of one was transplanted upside down onto the left half of the other, the right halves being transplanted similarly. We must assume that the two left and the two right halves then exerted strong inhibition one upon the other. Being both left or right "animals" they were the same genetically and therefore self-sufficient, though the reversed dorso-ventrality may also have played a significant role.

Serological factors in parabionts in general are discussed by Konieczna, Plonka, Skowron and Zabinski (1960) and Konieczna, Marczynska and Skowron-Cendrzak (1960).

This leads to the main problem of the widest significance in understanding the morphological diversities which depend on finely balanced interactions between induction and inhibition, both in ontogenesis and regeneration. In fact, the important problem of *pattern* is directly involved. We may therefore ask: how are induction and inhibition integrated with one another, and how are the regenerated parts determined so that they regenerate the required structures only?

It should again be stressed that we do not know what kind of biochemical agencies are released by injury. The expression "wound hormones" is convenient but has as yet no definite biochemical meaning. The events initiated after wounding are fairly well known, and are shown morphologically by a contraction of the wound with the expulsion of debris. The open wound is first covered by an incompletely defined and extremely thin amorphous layer, then by flattened, and finally normal, epidermal cells with lysis of the adjacent tissues. Neoblasts migrate to the wound where numerous mitoses occur, creating the blastema with sufficient differentiation exactly to form the missing parts. Why do they stop at forming only the missing parts and not go on to form whole new animals, in view of the well-established fact that neoblasts are totipotent? Several authors have maintained that the old part of the animal determines what is to be regenerated and think that this is correct, but what is the essential mechanism? Two main processes seem possible, one inductive and the other inhibitory.

In view of several instances of induction (eyes, pharynges, etc., discovered by the French school and others) emanating from the regenerating (embryonic) blastema itself, it seems improbable that adult tissues, except the nervous system, should induce and organize the differentiation of neoblasts in the blastema. To my mind it seems far more probable that inhibition is at work and I suggest the following mechanism:

The head, say, with a fairly large part of the body is removed by decapitation; the neoblasts in the blastema make a "high-point" by forming a brain and later, by induction from this brain, begin differentiation into eyes and other organs in due sequence, *but normally only in replacement of removed parts.* We know that neoblasts are able to build, say, a pharynx, and they could probably do so in the blastema formed after decapitation, but are prevented by the already existing pharynx, which releases inhibitory substances throughout the body (Chapter 10). The same applies to other organs still present in the adult body. I think that, in this matter, I am in agreement with the French school and others. My theory is therefore that differentiating neoblasts in the blastema sooner or later meet a biochemical inhibitory barrier from the adult body, resulting in the regeneration of an animal in a concerted manner.

The experiments discussed on p. 109 in my opinion support this theory. When the median parts, which are beginning to form eyes, are removed, the sides get a chance to form eyes themselves because the inhibition is abolished.

As another example it has been maintained by several authors that an open wound is a prerequisite for regeneration, but it is unnecessary for this wound to face externally. Figure 77 p. 92 presents such an experiment. A segment was removed and the remaining fore- and hind parts apposed with polarity maintained. The wound surfaces coalesced cleanly, and no blastema was visible externally. Nevertheless regeneration still took place, seemingly by the formation of an interior blastema, at any rate the newly interposed piece was at first unpigmented. The result was a normal animal, but of a smaller size than the original one. I think that this experiment is important because it shows without doubt that an external blastema—at least anteriorly—is not necessary to induce reorganization of missing parts. It also shows that incisions and missing parts can summon forces inside the body which are able to restore the equilibrium characteristic of the intact individual. One cannot help thinking of the rapid regeneration of mammalian liver (in rats and mice at least), when perhaps three-quarters of the liver has been removed; in such instances, in a few days normal equilibrium of the whole body is restored.

The experiment with *Bdellocephala* seems to indicate complete biochemical co-ordination in all parts of the individual. I have not yet had the opportunity to investigate the cellular events during this very interesting process of "inward" regeneration, but I am sure they should be studied. Such investigations might answer several fundamental questions: e.g. are neoblasts the only cells to build missing parts or do they play a minor role only? Do existing tissue cells in the two pieces transplanted simply duplicate themselves to restore the missing parts, as in liver regeneration, or does real morphallaxis take place? In any case biochemical forces to restore individuality must be involved, and inhibition must play a major role, because adult tissues do not permit the regenerate to exceed the growth of a normal animal.

This is not the place to develop the inhibition hypothesis further. I only wish to point out that, in my view, it may be extended to most morphogenetic and embryological processes, and that time is a very significant factor in inhibition.

2. *The nature of the time-graded regeneration field and its significance for heteromorphoses and polarity*

Chapter 5 dealt with the rates of regeneration throughout the body characteristic for every species. Regenerative processes start in the blastema at a rate determined by the region from which the blastema is formed. Discussion of the evidence was given by Brøndsted (1945, 1954, and 1956) and will now be amplified by more facts published since then.

It has been shown that the time-graded field cannot be ascribed to the number of neoblasts. The field is graded in a regular manner throughout the body (Figs. 103, 104) whilst the distribution of neoblasts is not. The metabolic rate expressed by the level of oxygenation, as held by Child and his school, cannot be the foundation either, because it has been shown that a similar gradient of respiration does not exist.

In view of the fact that the time-graded field is firmly laid down in the body, proven by the transplantation experiments by Brøndsted (p. 38–39, Fig. 29–33), we have to seek some morphological structure to which biochemical factors are tied, such factors being responsible for the different rates of regeneration throughout the body. My opinion is that only the nervous system conforms with these requirements. The significance of the nervous system has been discussed in Chapter 13. It has been shown by Lender and Klein (1961) and others that neurosecretory cells appear or are at least conspicuous when the animal is wounded, and it was suggested that neurohormones derived from them could be responsible for the migration and accumulation of neoblasts. Such a notion would be strengthened by a quantitative or semi-quantitative investigation of the number of neurosecretory cells along the entire length of the nervous system from the brain to the tailtip.

Even if these investigations should prove negative, the possibility still exists that the nervous system contains the biochemical factors for stabilizing the time-graded field. I suggest that we should investigate the content of neurocellular RNA along the nervous system as RNA accelerates regeneration (A. and H. V. Brøndsted, 1953). This could be done at least semi-quantitatively by histochemical and microspectrophotometric methods. The use of chemicals to inhibit the function of nerve cells might also be of some use.

It must be borne in mind that the bilateral distribution of the nervous system is relevant to lengthwise shifting of the time-graded field by transplantation as this also means shifting the two symmetrical halves of the nervous system, conforming with the idea of the nervous system as the morphological basis of the time-graded regeneration field.

If we adopt, as a working hypothesis, that the nervous system is biochemically responsible for phenomena concerning the time-graded field, it might be postulated that the nervous system in *Dendrocoelum* and *Bdellocephala* contains insufficient biochemical influence to start neoblasts building heads. We know that neoblasts assemble to form a blastema in these two genera after transverse incisions, both at the anterior and posterior surface of the cut, when it is made behind the pharynx, but at the anterior surface, the blastema seems to lack the impetus to start the differentiation necessary for head-building. On the other hand, a blastema formed at the posterior surface of the same transverse cut regenerates a tail quite readily.

How is this? One explanation could be that the biochemical impetus to form a tail is of a grade lower than that to form a head. This problem—as so many others in planarian regeneration—might be solved if a technique for culturing early blastema neoblasts could be devised and different parts of the nervous system were added to the culture.

Another explanation of the non-regeneration of a head from the posterior parts of the two genera would be that head-inhibitory substances were more abundant in the hind parts of these animals, although this seems to me a rather far-fetched idea.

In this connection we may note that several species and genera of Turbellaria are practically unable to regenerate. This fact cannot be due to "lower" organization because we find such genera both in "lower" and "higher" orders of Turbellaria. I think that it would be well worthwhile to investigate the biochemical properties of the nervous system in these non-regenerating genera and species.

It is obvious that the existence of a time-graded regeneration field bears directly on the phenemenon of polarity. The heteromorphoses, called Janus-heads and Janus-tails, represent complete reversals of polarity. These curious phenomena have been mentioned previously. Morgan and Child were aware of them. In most cases Janus-heads arise from very short transverse sections, though most readily from anterior sections, and Janus-tails most readily from extreme posterior sections. In both cases the very short sections need to be cut out from either a "strong" part of the time-graded field or from a very "weak" one. This may mean that the time difference between the biochemical events starting two blastemata is so short that the neoblasts of both begin differentiation by starting their "high-point" almost simultaneously, before inhibition becomes effective, resulting in a two-headed or two-tailed structure. This does not explain Janus-heads and -tails in long pieces. Flickinger (p. 50) and others have shown that pieces which contain considerable lengths of the time-graded field can reverse their polarity and form heteromorphoses, and Jenkins has found strains of animals with a strong tendency in this direction.

These phenomena are, of course, hard to explain in the light of the above assumption, and I think that we must fall back on explanations which include suppression of biochemical factors in normal animals.

Kanatani and Flickinger with co-workers (Chapter 6) produced a reversal of polarity by chemical methods. Kanatani tentatively ascribed this to oxidative alterations (cf. Child) whereas Flickinger was inclined to ascribe it to disturbances in protein synthesis and competition for available proteins. I am convinced that these outstanding experiments, when followed up and extended—preferably to species not propagating by fission—will greatly further our search for the basis of polarity, one of the great problems in the establishment of a biological morphological *pattern*.

Would it be unduly rash to connect the findings of Kanatani and Flickinger with the hypothesis of RNA in the nervous system as being *primus motor* for the existence of the time-graded regeneration field and therefore with the establishment of polarity and *pattern* in general? We know that cytoplasmic RNA is in some way coded by nuclear genetic DNA and that cytoplasmic RNA in some way codes synthesis. Would it be too far-fetched to imagine that in the "Janus" cases various chemicals might directly or indirectly influence the RNA in the nervous system so that its anterior–posterior and medio–lateral gradation was more or less suppressed? This would unbalance the normal gradation of protein synthesis, and hence provide the possibility of starting "high-points" in abnormal places flattening out the gradation of the time-graded field? Following this train of thought, would it not be reasonable to assume that some strains of planarians had inherent disturbances of gradation of nervous system RNA, e.g. the strains used by Jenkins?

Other examples of heteromorphoses can be discussed in the light of the tentative hypothesis concerning underlying causes for the existence of the time-graded regeneration field. Chapter 14 dealt with heteromorphoses of various kinds. Several authors have suggested that the nervous system is somehow involved, and in my opinion rightly so as most neoblasts in the intact animal are situated along the main nerve cords and we know that cuts call forth migration of neoblasts to the wound. When T-shaped wounds are made the blastemata build up their "high-points" in the "stronger" part of the time-graded field if the T-cuts are made anteriorly. This must give rise to new heads, even if the old head is present, because—as has been shown by Brøndsted (1954–6)—the process of head-production is so fast that inhibition by the old head will have had insufficient time to act, resulting in curious monsters, e.g. Fig. 88.

The same principle of time as a regulatory influence seems to be applicable to most other heteromorphoses, e.g. splitting experiments (Chapter 14), Randolph's experiment, formation of new polarities in side strips, extrusion of new heads after median splitting, formation of supernumerary eyes, pharynges and copulatory organs, and so on, and Runnström's conception of "Gefälleanarchie" may, in my opinion, be based on disturbances in the time-graded regeneration field.

3. *Morphallaxis, reindividualization and preservation of individuality during starvation and encystment*

It should be evident that the features enumerated in this section must have a common basis. In previous chapters most of them have been discussed thoroughly and it has been stressed that much work has still to be done to clarify what happens in the planarian body during these processes of cell development.

It is a well-known fact that planarians remain intact even after starvation severe enough to reduce them to 1 % of their original weight. The various organs become smaller unevenly, the sexual organs being the first to undergo regression. Berninger (1911), in *Dugesia gonocephala* and *D. alpina*, found that the animals shrank to one-twelfth of their original length ($\frac{1}{300}$ of original volume); although the eyes and sexual organs disappeared, only slight degeneration occurred in the parenchyma, intestine, nerves and muscles. Abeloos (1928) described a disproportionate decrease of organs during starvation. Only if the animal was reduced in length to about 1 mm was it no longer able to regenerate. The most hardy organs seemed to be the nervous system and parenchyma. It would be of the utmost interest to see how neoblasts behave during excessive starvation. My guess is that they would be amongst the last to succumb because we know that regeneration, even in severely starved animals, depends upon neoblasts and the nervous system.

In every case it is evident that during starvation morphallaxis must occur. Besides the problems discussed in Chapter 8, there remains this very important question: is the great decrease in body volume during starvation due to a decrease in cell numbers, cell size or both? Schultz (1904) was sure that the decrease was due only to a reduction in cell numbers and not cell size. Stoppenbrink (1905), on the contrary, ascribed the decrease in body volume to a decrease in cell volume and found considerable necrosis only in the sexual organs, which were the first to disappear. Other authors, e.g. Child (1915) and Abeloos and Lechamp (1929), found a decrease both in size and number of cells. I am at present engaged on this very problem. My preliminary results show that the volumes of the epithelial cell nuclei decrease considerably as do the numbers of cells, but curiously enough very little necrosis is seen. This brings me to the all-important question of how much do the amounts of RNA and DNA vary during starvation? We may assume that proteins, fat, carbohydrates and water decrease considerably in amount, therefore a decrease in the rate of regeneration subsequently may easily be understood, as pointed out on p. 153.

An important paper on starvation by Hoff-Jørgensen, Ebba Løvtrup and S. Løvtrup (1953) covers an investigation on *Polycelis nigra*. The animals were starved for 64 days during which time their fully extended length decreased from about 15 to 5 mm. From this it may be assumed that the volume decreased to about $\frac{1}{27}$. The authors found that total nitrogen decreased during the starvation period to about 14% of its initial value. In striking contrast to this the amount of DNA was almost the same after starvation as before, possibly slightly decreased. DNA, including all deoxyribonucleotides, was determined by growing *Thermobacterium acidophilus* in a medium containing those deoxyribonucleotides present in the experimental animals.

Although the technique used gave a reasonably accurate estimation of the DNA content, I would prefer reinvestigation by new biochemical methods. However, even if more accurate methods were to show greater losses of DNA, it seems beyond doubt that it is very stable even in severe starvation. We do not know, of course, how DNA and its breakdown products are distributed in starving animals, but in view of the fact that in this state they can regenerate complete individuals by morphallactic processes, the inference seems inescapable that the whole genetic code must be present in many cells, including neoblasts. Although we have no exact information as to the cellular morphology during encystment, we may tentatively infer that here also the genome must be linked to neoblasts.

Reindividualization may seem a strange subject in this section, but I think that this also must ultimately depend upon the same basic principles as the other subjects discussed here.

The morphological aspect of the problem was considered in Chapter 14. To get some idea as to how reindividualization is established I think that we have to resort to the fact, discussed in the first section of this chapter, that a bilateral animal is in reality two half-animals.

We have discussed several times how splitting the anterior half of a planarian causes each partial half to regenerate its own symmetrical portion, provided coalescence is prevented. When the two halves have completed regeneration the double-headed animal is connected only by its common hind part. When re-individualization starts, it begins at the junction just behind the two free fore parts, coalescence then gradually proceeding anteriorly until both heads are integrated; *pari passu* the right head loses its left part and the left head its right one.

I think that the most reasonable explanation of these events is that inhibitory forces from the right affect the left, and vice versa, in the intact animal to secure bilateral symmetry. From the undivided hind part of the animal these inhibitory forces gradually proceed forwards and so, presumably by serological influences, break down the artificially produced twinning in the fore parts. At the junction of unsplit and split parts of the animal inhibiting forces meet foreign tissue, where the undivided parts are in contact with the regenerated fore parts. The inhibiting parts of the undivided animal must therefore break down their counterparts whilst incompatability continues to exist, until the identical half-animals are restored to a single entity by coalescence of the two halves.

All other forms of re-individualization may be explained in the same way. This idea would also explain the many instances where, after longitudinal splitting, one divided part is later removed at or near the base of the split. When one split part is removed only a short way down, a head may regenerate probably indicating the existence of a time-graded field. The inhibitory forces from the undamaged part of the animal and the inductive mechanisms in the short or the long divided part compete with each other, the speed at which each operates determining the outcome.

In summary the essential fact is that both inhibition and induction in the planarian body operate to secure the unity (individuality) of an animal consisting of two symmetrical halves, which, although potentially individuals, normally blend to form the whole animal.

The harmonious integration of the two mechanisms sometimes breaks down, as in the numerous examples of tumour-like protuberances which occur. Tumours in planarians are very interesting, but I think that our knowledge is still so scanty that I shall refrain from discussing them in this book.

4. Conclusion

We have now presented an exhaustive body of evidence and hypotheses about planarian regeneration. Tentative hypotheses based on the salient facts have been proposed in this chapter, but the core of the regenerative problem is still genetic. Chapter 21 dealt with some of the essentially basic problems. In the absence of adequate work on the genetic side, I can here only theorize briefly.

Before doing so it would seem reasonable to recapitulate the outstanding features of planarian regeneration.

(a) A wound-stimulus in all probability calls forth wound hormones, i.e. humoral forces of unknown nature which elicit some re-

markable features in the injured body, namely: histolysis in
the first $\frac{1}{2}$ mm near the wound but often penetrating deeper.

(b) Products of lysis presumably serve as nourishment.

(c) Phagocytosis, rearrangement and destruction of tissue by morphallaxis occurs (Chapter 8 where the important feature of the disappearance of fibrous tissue is mentioned).

(d) Migration and mitosis of neoblasts (Chapters 7, 8, 11, and 12).

(e) Stimulation of the nervous system as *primus motor* in the regenerative
process (Chapter 13).

(f) Induction of organs with "organisin", a substance not specific to any
species (Chapter 9).

(g) Inhibitory phenomena involving specific substances (Chapter 10).

(h) Time as an overall influence in regenerative processes (Chapters 5,
17).

(i) Stability of the time-graded regenerative field (Chapter 5).

(j) Restoration of the time-graded field.

(k) Restoration of individuality (Chapter 13).

(l) Biochemical and metabolic mechanisms associated with the synthesis
of proteins and enzymes (Chapters 4, 17).

(m) Influence of size, age, temperature (Chapter 15) and starvation (Chapter 23, Sect. 3).

Chapter 21 concluded with the words: "It is therefore clear that a blueprint for pattern must exist. It is our task to find out how this blueprint is
organized and how it works."

Let us imagine an extreme situation. A small part is removed from the
side of an animal, say as in Fig. 24, without including any portion of the
main nerve cord. This little piece can nevertheless transform itself into a small
but complete worm with brain, eyes (the right one first), all organs and so on.
We know that neoblasts migrate to the wounds and all available evidence
suggests that they are totipotent. We have evidence that the nerve-endings in
the parts cut out call upon the neoblasts to migrate to the wound, where we
have a collection of neoblasts, all totipotent. Now how can the neoblasts
release their genetic code in such a way that a properly proportioned animal
develops? The neoblasts are identical genetically so why should only some of
them begin to regenerate a brain first? Why do they not all begin to make the
same organ, thus transforming the piece cut out into just one large brain?

Remembering the pessimistic comments of Holzinger (pp. 210 and 212),
I have nevertheless attempted to form a hypothesis of my own as follows
(1954):

In the following working hypothesis, the differentiation of the blastema is interpreted in terms of successive gene actions and inhibitions exercised by the median

parts over the more peripheral parts of the blastema. Centrally placed neoblasts (Fig. 24) differentiate more quickly, that is to say, their cytoplasm more quickly reaches such a state as to respond to brain-determining gene actions. In differentiating they inhibit the neighbouring cluster of cells (group 2) from responding to brain-forming gene-actions. The next gene action in time is, e.g., eye-determining, and to these adjacent cells now respond. Neoblasts later in the order of differentiation are inhibited from forming brain and eyes, and so form muscle (group 3), intestine (group 4), or do not differentiate (group 5). These last remain totipotent.

Although this proposition is somewhat naïve, nevertheless I think it of some use. Flickinger (1962) developed a more modern approach in his article and I strongly recommend it to those interested in this basic problem.

In all events I cannot see how we can omit time as a factor in triggering off gene actions during the regeneration of a separated small part into a whole animal. Nor can I see how we can escape the acceptance of there being some sort of interaction between cytoplasm and genes in the time-regulated differentiating neoblasts, which build up the various kinds of tissues during regeneration.

It is tempting here to describe recent ideas concerning cytoplasmic influences on the gene, such as the operon theories developed by Jacob and Monod and others; see, for example, W. Braun, *Bacterial Genetics*, 1965. This is not the place, however, to discuss these theories, apart from mentioning repressor-genes, which may be involved in the suppression of certain gene-actions occurring after the first messenger RNA has been fabricated. They inhibit continued full action of the same gene, but at the same time let other parts of the gene function, releasing other messenger RNA which leads to new proteins characteristic of other cell types. Time would be involved in these procedures and so be the decisive influence in determining new cell types leading to fresh differentiations.

All this is, of course, pure speculation, but I think that a meticulous histochemical investigation into the time sequence of the development of cell types in the blastema might be useful.

In elucidating the essentials of regeneration in planarians it is necessary in all cases to take time into consideration quantitatively. Such a study might help explain the existence of the time-graded regeneration field: if, as we have good reason to believe, some substance from the nervous system activates first the cytoplasm of neoblasts and then the genome, then it might be readily understood why the stronger the activation the sooner the response of the genome. The more concentrated the activating substance of nervous origin the sooner would a "high-point" be reached. Such an idea would conform with the anatomical arrangement of the nervous system, and afford one more explanation for the time-graded regeneration field. A very important qualification of such an idea would be that such a mechanism would not work if there was inhibition from neighbouring parts of the body, e.g. if the cut was situated so that it faced posteriorly. In such a situation a tail high-point should ensue, which is borne out by experiment.

I think it only fair and necessary to conclude this book by saying that planarian regeneration still has many unsolved problems presenting a wide experimental field, which should prompt scientists to use all their skill to develop new techniques for solving them. I sincerely hope that this book—despite its obvious shortcomings—will be of help to future investigators.

ADDENDUM

AFTER I had completed my manuscript, Professor Etienne Wolff kindly sent me two manuscripts of important work from his institute with permission to incorporate them in my references, although unpublished. I thank the authors for their courtesy.

Some of their contents have been given in a preliminary form and mentioned briefly in this book (p. 93, Fig. 78) but the investigations have since been enlarged and confirmed to such a degree that I am glad I can quote their results in this addendum.

The first, written by Catherine Ziller-Sengel, is *L'Inhibition spécifique de la régénération du pharynx chez les planaires.* Two methods were used, transplantation and making filtered brei. By the first, a supplementary pharynx *(Dugesia lugubris)* was transplanted anteriorly to the normal one in sixty-five animals (Fig. 160 shows the procedure), but with three variations in the

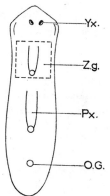

FIG. 160. *Dugesia lugubris.* See text. (From Ziller-Sengel, MS. 1966.) Yx, eyes; Zg, zone of transplanted pharynx; Px, host pharynx; OG, genital orifice.

procedure. (i) In forty the normal pharynx was then removed; all these animals regenerated a new pharynx in the usual time, 5–9 days. (ii) In seventeen the transplanted pharynx was subsequently removed, leaving the normal one *in situ*. In seven of these no additional pharynx regenerated. Four of the other ten regenerated an additional pharynx anterior to the old one, but more slowly than group (i) (16–17 days). (iii) In eight both the old and the newly transplanted pharynx were removed; in all cases a new pharynx was regenerated at the normal site in the host. The author concluded that the old pharynx either completely or partially inhibited regeneration of a new pharynx.

We know that the pharynx has a surrounding area of pharyngeal inductive tissue. The author's conclusions appeared to be that even a transplanted pharynx created such an area which could sometimes repel inhibition from the old one. Figure 161 shows this idea schematically.

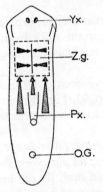

Fig. 161. *Dugesia lugubris*. Diagram see text. (From Ziller-Sengel, MS. 1966.) Yx, eyes; Zg, zone of transplanted pharynx; Px, host pharynx; OG, genital orifice.

Figure 162 shows another interesting experiment. In twenty-five animals a fresh pharynx with some pharynx-inducing area was transplanted beside the normal pharynx with the same polarity.

(i) Then in sixteen of the twenty-five the normal pharynx of the host was excised; in three specimens of this latter group the transplanted pharynx took the place and functions of the old one without regeneration of a new one, but in two the transplanted pharynx simply remained at the transplantation place again without regeneration of a new one; in the remaining eleven a new pharynx regenerated from the site of the old though slowly in two cases.

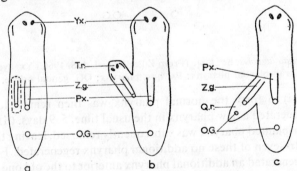

Fig. 162. *Dugesia lugubris*. Diagram see text. (From Ziller-Sengel, MS. 1966.) Qr, regenerated tail; Tr, regenerated head.

(ii) In six of the twenty-five the transplanted pharynx was removed again. One of these failed to grow an additional pharynx possibly due to the fact that in this experiment the first pharynx had been transplanted very near the old one, enabling the latter to inhibit regeneration of the excised transplant. The other five of this group healed poorly because the transplants had been placed very laterally but all five regenerated new pharynges, according to the author because their fresh pharynges had been out of reach of the inhibiting force of the old pharynx. An important point emerged that new heads also regenerated from new pharynges if the latter protruded anteriorly from the host's body, and new tails regenerated if they protruded posteriorly. This seems to indicate that neoblasts assembled in the neighbourhood of the open wound and regenerated according to the polarity of the pharyngeal zone exposed.

The second method used by the author involved making brei and extracts thereof. The procedures should be read in the original as only a short summary of these very interesting experiments will be given. They can be divided into three major groups:

1. In the first group the pharynx with the surrounding zone was removed in all, and they were then divided into five sub-groups. The animals were kept in a culture media containing (i) crude brei from the head region; (ii) crude brei from the tail region; (iii) brei from the pharynges alone; (iv) brei from the whole pharyngeal region; (v) in pure water (controls).
2. Animals with their pharynges excised were put in *filtered* solutions from the same regions as in group 1.
3. As in group 2, but put in identical filtered solutions, several days after pharyngectomy, i.e. when the regeneration of new pharynges had already begun.

Group 1 (Fig. 163) showed some inhibition of pharyngeal regeneration compared with the controls in pure water. However, the author ascribed this

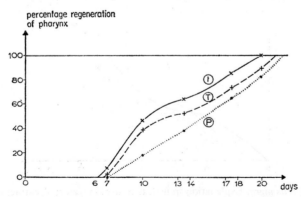

FIG. 163. *Dugesia lugubris*. See text. (From Ziller-Sengel, MS. 1966.) P, extract of pharyngeal region; T, extract of heads; t, control in water.

to bacterial contamination only. Group 2 (*Polycelis nigra*, Fig. 164) showed a considerable inhibition from pharynx + pharyngeal area and of pharyngeal area + pharyngeal extracts. It should be noted that only extracts from the pharyngeal area inhibited strongly.

In group 3 the removal of the pharynx was made only after 3 days after the addition of the extracts to the culture medium. These results are of great

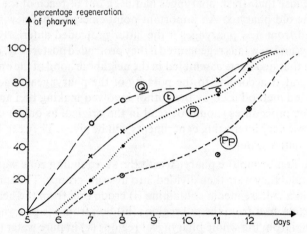

FIG. 164. *Dugesia tigrina*. Explanation in text. Q, extract of tail; t, control in water; P, extract of total pharyngeal region; Pp, extract of pharyngeal region without pharynx. (From Ziller-Sengel, MS. 1966.)

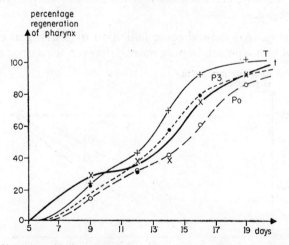

FIG. 165. *Polycelis nigra*. Explanation in text. t, control in water; T, extract from heads; Po, extract from pharyngeal zone without pharynx; P3, extract from pharyngeal zone without pharynx after 3 days. (From Ziller-Sengel, MS. 1966.)

importance because they show that once pharyngeal regeneration has begun, it develops partial or complete insensitivity to the inhibitory influence of the extracts. Figure 165 shows the growing insensitivity towards substances similar to those formed in the regenerating organ. The experiments are very satisfying because they agree with the experiments of Brøndsted (1954) showing that a regenerating head inhibits new heads from parts lateral to the "high-point" (Chapters 5 and 23).

The paper of Ziller-Sengel confirms another important point, namely that *Dugesia tigrina*, a species propagating by fission, is unsuitable for these experiments due to the presence of zooids. As before emphasized, experiments on planarians with asexual propagation *must not* be applied without reserve to those with only sexual means of propagation. The paper of Ziller-Sengel adds materially to the subjects discussed in Chapters 5 and 9.

The second paper by Barbara Fedecka Bruner was entitled: «Etudes sur la régénération des organes géniteaux chez la planaire *Dugesia lugubris*.»

A short summary of the author's preliminary account dealing with restitution of testicles after irradiation was given on p. 100. In her thesis these investigations were enlarged with further experimental evidence especially on the regeneration of copulatory organs. Several important theoretical points emerged and are fully reviewed below.

The first section (pp. 1–61 in the MS.) dealt with the testicles, the second (pp. 62–112) with regeneration of the copulatory organs. The author first dealt fully with the literature on gonads in planarians.

The testicles in the young animal are formed by neoblasts. Mature testicles are encapsulated by spermatogones ("une sorte d'epithelium germinatif") surrounding different stages of spermatozoon-formation. In *Dugesia lugubris* they are scattered dorsally. This fact is utilized for the planning and execution of her experiments, which involved either surgery or irradiation.

When the anterior third of the animal was cut away, the remaining testicles degenerated and after 5 days: «... les testicules ne sont plus représentés que par des débris épars dans le parenchyme». After this new testicles began to appear.

When either the posterior part of the animal just behind the pharynx, the pharynx and copulatory organs or the two nerve cords were removed the same phenomenon was observed. However, when only lateral strips of the animal were cut away, the testicles remained normal. If just the pharynx was removed then only the testicles adjacent to the wound degenerated whilst all other testicles remained quite normal. The same applied if the copulatory organs only were removed.

If one lateral nerve cord was removed, the testicles degenerated and no regeneration took place subsequently. When the brain was removed the testicles anterior to the pharynx degenerated. The results are summarized on Fig. 166.

The author concluded that testicular degeneration depended primarily on how much of the body was excised. Removal of one-third resulted in complete degeneration as did the removal of one-sixth, but then only if the operation involved the removal of nerve cords. Removal of the brain, which is only a minute part of the body, had a profound influence on their degeneration.

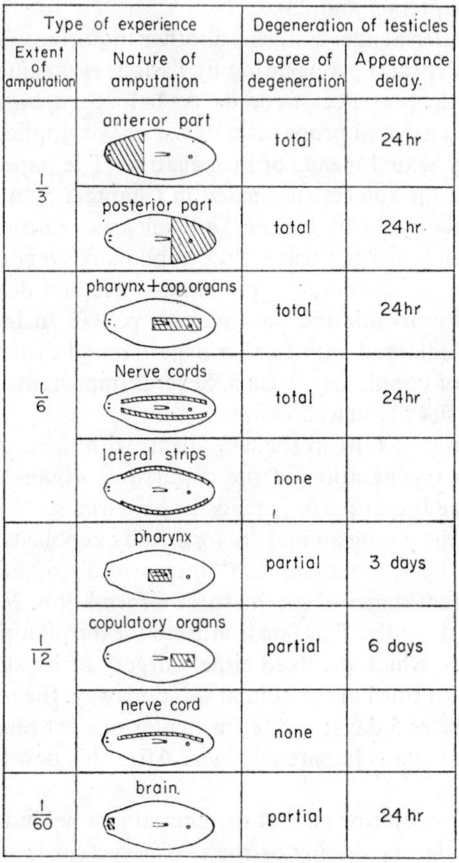

Type of experience		Degeneration of testicles	
Extent of amputation	Nature of amputation	Degree of degeneration	Appearance delay.
1/3	anterior part	total	24 hr
	posterior part	total	24 hr
1/6	pharynx+cop.organs	total	24 hr
	Nerve cords	total	24 hr
	lateral strips	none	
1/12	pharynx	partial	3 days
	copulatory organs	partial	6 days
	nerve cord	none	
1/60	brain.	partial	24 hr

FIG. 166. *Dugesia lugubris*. Diagram of operations. Explanation in text. (From Fedeckaa Bruner, MS. 1966.)

The author drew three important conclusions:

1. The gonadal cells of the testicles seemed to be very like neoblasts. After testicular degeneration, these cells alone seemed to survive, migrated into the parenchyma, where they were indistinguishable from neoblasts, and probably took part in the regenerative processes as ordinary neoblasts.

2. The nervous system was necessary for the maintenance of testicles.

3. A hierarchy of organ formation seemed to exist, the nervous system being dominant and the testicles (and ovaries?) subordinate. This idea is strengthened by the fact that neurosecretory phenomena are concerned with the development of sexual maturity.

This author's experiments with X-rays which were briefly mentioned on p. 100 are given in detail in her new paper.

In the experiments concerning the copulatory organs, she first pointed out that they appeared late in ontogenesis, and were very labile, that is, apt to degenerate both under natural and experimental conditions. They might disappear completely during seasonal periods, or after prolonged starvation, but they also regenerated rapidly. The author was concerned with how their regeneration took place. After a clear account of the various hypotheses set forth in the literature, and a full description of the organs, she then reported her experiments on them.

Normal regeneration after removal was rapid from January to May, which is the egg-laying period. In July the organs did not regenerate at all, but early in the autumn regeneration again took place. Therefore she chose to do all her experiments in the period from January to May.

The first set were limited to removing the organs proper with a meticulous description of the morphallactic regeneration process leading to their regeneration, which was complete after 25–30 days.

After these preliminary investigations the author carried out a large series of experiments to determine if regeneration of the copulatory organs was dependent on the amount of tissue removed. She also irradiated the testicles. She concluded that testicles were prerequisite for regeneration of copulatory organs. A certain minimum of neoblasts and at least some part of the nervous system were also found necessary. The removal of part of the nerve cord therefore influenced the regeneration of the copulatory organs, a fact which the author phrased thus: «Dans tous les cas, l'appareil copulateur ne se raconstitue que lorsque tous les autres organes de la planaire sont régénérés ou en voie de régénération avancée.»

The essence of her work was that *induction* from the testicles occurs. We may therefore again emphasize the important finding of the Wolffian school that there is a hierarchy of induction in planarian regeneration.

REFERENCES

ABELOOS, M. (1927a) La vitesse de régénération de la tête chez *Planaria gonocephala*. *C.R.Acad.Sc. Paris* **184,** 345.

ABELOOS, M. (1927b) Sur la perte de poids des Planaires (*Pl. gonocephala* Dugès) en régénération à différentes températures. *C. R. Soc. Biol.* **96,** 925–6.

ABELOOS, M. (1927c) La vitesse de régénération de la tête chez *Planaria gonocephala* Dugès. Influence de la taille des individues opérés. *C.R. Soc. Biol.* **96,** 1923–4.

ABELOOS, M. (1927d) Les théories de la polarité dans les phénomènes de régénération. *Biol. Rev.* 91–128.

ABELOOS, M. (1927e) Les corrélations organiques et les théories de C. M. Child. *Rev. Gén. des Sc.* **37,** 395.

ABELOOS, M. (1928) Sur la dysharmonie de croissance chez *Planaria gonocephala* Dugès et sa réversibilité au cours du jeune. *C.R. Soc. Biol.* **98,** 917–19.

ABELOOS, M. (1929) Influence de la température sur la croissance des Planaires. *C.R. Acad. Sc. Paris* **188,** 881–3.

ABELOOS, M. (1930) Recherches expérimentales sur la croissance et la régénération chez les Planaires. *Bull. Biol. France et Belg.* **64,** 1–140.

ABELOOS, M. (1932) *La Régénération et les Problèmes de la Morphogénèse*. Gauthier-Villars, Paris.

ABELOOS, M. (1954) Régénération et différentiation cellulaire. *Ann. de l'Université (Poitiers)*.

ABELOOS, M. (1965) La régénération des Annelides. In *Regeneration in Animals*, Amsterdam, pp. 207–15.

ABELOOS, M. and LECAMP, M. (1929) Sur la taille des cellules épithéliales tégumentaires au cours de la croissance et au cours du jeune chez les Planaires. *C. R. Soc. Biol.* **101,** 899.

AISUPET, M.P. (1937) Der Einfluß des Regenerationsprozesses auf den Organismus. *Russ. Biol. J.* **6.**

ALEXANDER, M.J. and PRICE, H.F. (1926–7) Histological degeneration preceding fragmentation and encystment in *Planaria velata*. *Anat. Rec.* 34, Abstr. 151.

ALLEN, G.D. (1919) Quantitative studies of the rate of respiratory metabolism in Planaria. I. The effect of potassium cyanide on the rate of oxygen consumption. *Amer. J. Phys.* **48,** 93–120. II. The rate of oxygen consumption during starvation, feeding, growth and regeneration in relation to the method of susceptibility to potassium cyanide as a measure of rate of metabolism. *Ibid.* **49,** 420–73.

ALLEN, G.D. (1919) The rate of carbon dioxide production in pieces of *Planaria* in relation to the theory of metabolic gradients. *Proc. Amer. Soc. Zool. Anat. Rec.* **17,** 1920.

ANDERSON, LOUISE A. (1927) The effect of alkalis on the oxygen consumption and susceptibility of *Planaria dorotocephala*. *Biol. Bull.* **53,** 327–42.

ANSEVIN, K.D. and BUSCHBAUM, R. (1961) Observations on planarian cells cultivated in solid and liquid media. *J. Exp. Zool.* **146,** 153–61.

ASPEREN, K. V. (1946) Pharynx regeneration in postpharyngeal fragments of *Polycelis nigra*. *Proc. Ned. Ak. Wet.* **49,** 1083–90.

AUTUORI, I., NARDELLI, M.B., and GABRIEL, A. (1965) Les enzymes protéolytique et les phosphatases au cours de la régénération de fragments Planaires. *C. R. Acad. Sc. Paris* 995-8.

AX, PETER (1957a) Vervielfachung des männlichen Kopulationsapparates bei Turbellarien. *Verh. d. Zool. Ges. in Graz.*

AX, PETER (1957b) Ein chordioides Stützorgan des Entoderms bei Turbellarien. *Z. Morph. u. Ökol. d. Tiere.* **46**, 389-96.

AX, PETER (1958) Verwandtschaftsbeziehungen und Phylogenie der Turbellarien. *Ergeb. Biol.* **24**, 1-68.

AX, PETER and SCHULTZ, E. (1959) Ungeschlechtliche Fortpflanzung durch Paratomie bei acoelen Turbellarien. *Biol. Zentr. Bl.* **78**, 613-22.

BAHRS, ALICE M. (1931) The modification of the normal growth-promoting power, for planarian worms, of the digestive mucosa of the rabbit under variations in diet, fasting and age. *Phys. Zool.* **4**, 189-203.

BAHRS, A.M. and WULZEN, R. (1932) Variations in the growth-promoting power of kidney for planarian worms. *Phys. Zool.* **5**, 192-206.

BANDIER, J. (1936) Histologische Untersuchungen über die Regeneration von Landplanarien. *Arch. Entw. Mech.* **135**, 316-48.

BARDEEN, C.R. (1901) On the physiology of the *Planaria maculata* with special reference to the phenomenon of regeneration. *Amer. J. Phys.* **5**, 1-55.

BARDEEN, C.R. (1902) Embryonic and regenerative development in Planarians. *Biol. Bull.* **3**, 262-88.

BARDEEN, C.R. (1903) Factors in heteromorphosis in planarians. *Arch. Entw. Mech.* **16**, 1-20.

BARDEEN, C.R. and BAETJER, F.H. (1904) The inhibiting action of the roentgen rays on regeneration in planarians. *J. Exp. Zool.* **1**, 191-6.

BARTSCH, O. (1923) Die Histiogenese der Planarienregenerate. *Zool. Anz.* **56**, 63-67.

BARTSCH, O. (1923b) Die Histiogenese der Planarienregenerate. *Arch. Mikr. Anat. u. Entw. Mech.* **99**, 187-221.

BECKER, R.O. (1961) The bioelectric factors in amphibian-limb regeneration. *J. Bone and Joint Surg.* **43** A(5), 643-56.

BEHRE, E.H. (1918) An experimental study of acclimatisation to temperature in *Planaria dorotocephala. Biol. Bull.* **35**, 277-318.

BEISSENHIRTZ, H. (1928) Experimentelle Erzeugung von Mehrfachbildungen bei Planarien. *Z. wiss. Zool.* **132**, 257-313.

BELKIN, R.I. (1947) Les hormones de blessure, leur formation, leur importance pour la régénération (en russe). *Sovietsk. Biol.* 61-88.

BELLINI, L., CECERE, F., and ZAPPANICO, AGNESE (1962) Incorporazione di aminoacidi marcati e comportamente delle proteasi nella regenerazione di *Dugesia lugubris. Red. Inst. Sc., Univ. Camerini*, 252-60.

BELLINI, L. and POLZONETTI, A.M. (1962) *Rend. Inst. Sc. Camerino* **3**, 201-10.

BENAZZI, M. (1942) Risultati di ricerche genetiche sulle Planarie. *Soc. Ital. Biol. Sper.* **17**, 1-2.

BENAZZI, M. (1942b) Ricerche genetiche sulla scissiparita di *Dugesia gonocephala. Arch. Ital. di Anat. e di Emb.* **47**, 72-94.

BENAZZI, M. (1966) Considerations on the neoblasts of planarians on the basis of certain karyological evidence. *Chromosoma* **19**, 14-27.

BERGENDAHL, D. (1892) Studien über Turbellarien. 1. Über die Vermehrung durch Querteilung des *Bipalium kewense. Mos. Sv. Vet. Ak. Handl.* **25**, 1-40.

BERNINGER, J. (1911) Über die Einwirkung des Hungers auf Planarien. *Zool. Jahrb. Abt. f. Zool. u. Phys.* **30**, 181-216.

BERRILL, N.J. (1952) Regeneration and budding in worms. *Biol. Rev.* **27**, 401-38.

BERRILL, N.J. (1961) *Growth, Development and Pattern*, San Francisco and London. 555 pp.

BERTALANFFY, L. v. (1942) Wachstumgradienten und metabolische Gradienten bei Planarien. *Biol. General.* **15**, 295–311.

BERTALANFFY, L. v. (1952) Planarians as model organisms for morphogenesis and pharmacodynamical action. *Rev. Canad. Biol.* **11**, 54.

BERTALANFFY, L. v., HOFFMANN, O., and SCHREIER, O. (1946) A quantitative study of the toxic action of some quinones on *Planaria gonocephala*. *Nature* **158**, 498.

BETCHAKU, T. (1960) A copper sulphate-silver nitrate method for nerve fibers of planarians. *Stain Tech.* **35**, 215–18.

BEYER, K.M. and CHILD, C.M. (1933) Reconstitution of lateral pieces of *Planaria dorotocephala* and *Planaria maculata*. *Phys. Zool.* **3**, 342–65.

BLUMBERG, TH.T. (1940) Regeneration of *Euplanaria dorotocephala* with pituitary gland extract. *Proc. Soc. Exper. Biol. Med.* **44**, 117–19.

BÖHMIG, L. (1909) Planarien in *Süßwasser Deutschlands*, Heft 19.

BÖHMIG, L. (1913) Studien an Doppelplanarien. Die Kokonbildung und -ablage bei Planarien mit vermehrter Zahl der Copulationsapparate. *Zool. Jahr. Abt. Anat. Ontol. d. Tiere.* **36**, 307, 336.

BONDI, C. (1959) Osservazioni sui rapporti tra rigenerazione degli occhi e sistema nervoso in *Dugesia lugubris*. *Arch. Zool. Ital.* **44**, 141–50.

BONNER, J.T. (1963) Epigenetic development in the cellular slime moulds. In *Symposia of the Soc. Exper. Biol.* **17**, *Cell Differentiation*, pp. 341–58.

BONNER, J.T. and HOFFMAN, MARY E. (1963) Evidence for a substance responsible for the spacing pattern of aggregation and fruiting in the cellular slime molds. *J. Embr. Exp. Morph.* **11**, 571–89.

BRACHET, J. (1946) Aspects biochimiques de la régénération. *Experientia* **2**, 41–48.

BRACHET, J. (1950) *Chemical Embryology*, London.

BRACHET, J. (1960a) The role of sulfhydril groups in morphogenesis. *J. Exper. Zool.* **142**, 115–39.

BRACHET, J. (1960b) *The Biological Role of Ribonucleic Acids*, Elsevier Publishing Co.

BROWN, F. (1962a) Response of the planarian, *Dugesia*, and the protozoan *Paramecium*, to very weak horizontal magnetic fields. *Biol. Bull.* **123**, 264–81.

BROWN, F. (1962b) Response of the planarian, *Dugesia*, to very weak horizontal electrostatic fields. *Biol. Bull.* **123**, 282–94.

BROWN, F. (1963) An oriental response to weak gamma radiation. *Biol. Bull.* **125**, 206–25.

BROWN, F.A., JR. and PARK, T.L.H. (1964) Seasonal variations in sign and strength of gamma-taxis in planarians. *Nature* **202**, 469–71.

BRØNDSTED, H.V. (1937) Experiments with methylene blue on the reducing capacity of *Dendrocoelum*. *Protoplasma* **27**, 556–62.

BRØNDSTED, H.V. (1939) Regeneration in planarians investigated with a new transplantation technique. *Kgl. D. Vid. Selsk. Biol. Medd.* **15**, 1–39.

BRØNDSTED, H.V. (1942a) Further experiments on regeneration-problems in planarians. *Kgl. D. Vid. Selsk. Biol. Medd.* **17**, 1–27.

BRØNDSTED, H.V. (1942b) A study in quantitative regeneration: the regeneration of eyes in the planarian *Polycelis nigra*. *Vid. Medd. D. Naturh. Foren.* **106**, 253–62.

BRØNDSTED, H.V. (1942c) Experiments with LiCl on the regeneration of planarians. *Ark. f. Zool.* **34**, B. 1–7.

BRØNDSTED, H.V. (1946) The existence of a static, potential and graded regeneration field in planarians. *Kgl. D. Vid. Selsk. Biol. Medd.* **20**, 1–31.

BRØNDSTED, H.V. (1947) The time-graded regeneration field in planarians. *Proc. 6th Intern. Congr. Exper. Cytol.*, pp. 585–7.

BRØNDSTED, H. V. (1953) Rate of regeneration in planarians after starvation. *J. Embr. Exp. Morph.* **1**, 43–47.

BRØNDSTED, H. V. (1954) The time-graded regeneration field in planarians and some of its cyto-physiological implications. *Colston Papers* **7**, 121–38.

BRØNDSTED, H.V. (1955) Planarian regeneration. *Biol. Rev.* **30**, 65–126.

BRØNDSTED, H.V. (1956) Experiments on the time-graded regeneration field in planarians. *Kgl. D. Vid. Selsk. Biol. Medd.* 3–39.

BRØNDSTED, H.V. (1962) Entwicklungsphysiologie der Porifera. *Fortschr. d. Zool.* **14**, 115–29.

BRØNDSTED, AGNES and H.V. (1952) The time-graded regeneration field in *Planaria (Dugesia) lugubris*. *Vid. Medd. D. Naturh. Foren.* **114**, 443–7.

BRØNDSTED, AGNES and H.V. (1953) The acceleration of regeneration in starved planarians by ribonucleic acid. *J. Embr. Exp. Morph.* **1**, 49–54.

BRØNDSTED, AGNES and H.V. (1954) Size of fragment and rate of regeneration in planarians. *J. Embr. Exp. Morph.* **2**, 49–52.

BRØNDSTED, AGNES and H.V. (1961a) Influence of temperature on rate of regeneration in the time-graded regeneration field in planarians. *J. Embr. Exp. Morph.* **9**, 159–66.

BRØNDSTED, AGNES and H.V. (1961b) Number of neoblasts in the intact body of *Euplanaria torva* and *Dendrocoelum lacteum*. *J. Embr. Exp. Morph.* **9**, 167–72.

BRUNST, V.V. (1961) Some problems of regeneration. *Quart. Rev. Biol.* **36**, 178–206.

BUCHANAN, J.W. (1922) The control of head form in Planaria by means of anesthetics. *J. Exp. Zool.* **36**, 1–47.

BUCHANAN, J.W. (1923) On the nature of the determining factors in regeneration. *J. Exp. Zool.* **37**, 395–416.

BUCHANAN, J.W. (1926a) Depression of oxidative metabolism and recovery from dilute potassium cyanide. *J. Exp. Zool.* **44**, 285–306.

BUCHANAN, J.W. (1926b) Some antagonistic and additive effects of anesthetics and potassium cyanide. *J. Exp. Zool.* **44**, 307–25.

BUCHANAN, J.W. (1927) The spatial relations between developing structures. I. The position of the mouth in regenerating pieces of *Phagocata gracilis* (Leidy). *J. Exp. Zool.* **49**, 69–92.

BUCHANAN, J.W. (1930) The nature of disintegration gradients. I. The significance of a gradient in susceptibility to distilled water in planaria. *J. Exp. Zool.* 307–30.

BUCHANAN, J.W. (1930b) II. The effect of hypertonic solutions on the disintegration of planaria by high temperatures. *J. Exp. Zool.* 455–72.

BUCHANAN, J.W. (1933) Regeneration in *Phagocata gracilis* (Leidy). *Phys. Zool.* **6**, 185–204.

BUCHANAN, J.W. (1934–5) An analysis of physiological states responsible for antero-posterior disintegration in *Planaria dorotocephala*. *Protoplasma* **22**, 497–512.

BUCHANAN, J.W. (1938) The effect of planarian extract and exudates upon head regeneration in *Euplanaria dorotocephala*. *Phys. Zool.* 144–54.

BUCHANAN, J.W. and LEVENGOOD, C.A. (1939) Axial susceptibility to serum antibodies in *Euplanaria*. *Phys. Zool.* 56–71.

CARRIERES, J. (1882) Die Augen von *Planaria polychroa* Schmidt und *Polycelis nigra* Ehrb. *Arch. f. Mikr. Anat.* **20**, 160–72.

CASTLE, W.A. (1927) The life history of *Planaria velata*. *Biol. Bull.* **53**, 139–44.

CASTLE, W.A. (1928) An experimental and histological study of the life-cycle of *Planaria velata*. *J. Exp. Zool.* **51**, 417–83.

CASTLE, W.A. (1940) Methods for evaluation of head types in planarians. *Phys. Zool.* **13**, 309–33.

CECERE, F., GRASSO, M., URBANI, E., and VANNINI, E. (1964) Osservazioni autoradiografiche sulla rigenerazione di *Dugesia lugubris*. *Rendic. Inst. Sc. Univ. Camerino*.

CHALKLEY, D.T. (1959) The cellular basis of limb regeneration. In *Regeneration in Vertebrates* (ed. C.S.THORNTON), Univ. Chicago Press, Chicago, pp. 34–58.

CHANDEBOIS, ROSINE (1950) Inhibition partielle de la régénération chez la planaire marine *Procerodes lobata* O. Schm. *C. R. Séanc. Acad. Sc.* **231**, 1347–8.

CHANDEBOIS, ROSINE (1951) Etude expérimentale des régénérations asymétriques chez la planaire marine *Procerodes lobata* O. Schm. *Bull. Soc. Zool. Fr.* **76**, 404–8.

CHANDEBOIS, ROSINE (1952a) Hétéromorphoses et hémihétéromorphoses chez la planaire marine *Procerodes lobata* O. Schm. *C. R. Séanc. Acad. Sc.* **234**, 1319–21.

CHANDEBOIS, ROSINE (1952b) Variations régionales des processus de cicatrisation chez la planaire marine *Procerodes lobata* O. Schm. *C. R. Séanc. Soc. Biol.* **146**, 1950.

CHANDEBOIS, ROSINE (1953) Déterminisme des anomalies de la régénération de la tête chez la planaire marine *Procerodes lobata* O. Schm. *C.R. Séanc. Acad. Sc.* **236**, 330–2.

CHANDEBOIS, ROSINE (1954) Comportement de l'épiderme dans les régénérats tératomorphiques chez la planaire marine *Procerodes lobata* O. Schm. *C.R.Séanc. Acad.Sc.* **239**, 911–13.

CHANDEBOIS, ROSINE (1955) Sur les phénomènes d'hémihétéromorphose chez la planaire marine *Procerodes lobata* O. Schm. *Bull. Soc. Zool. Fr.* **80**, 139.

CHANDEBOIS, ROSINE (1957) Recherches expérimentales sur la régénération de la planaire marine *Procerodes lobata* O. Schm. *Bull. Biol.* **91**, 1–94.

CHANDEBOIS, ROSINE (1960) Sur la source de l'histogenèse régénératrice chez les planaires. *C. R. Acad. Sc.* **251**, 146–8.

CHANDEBOIS, ROSINE (1962) Role des éléments fixes et libres du parenchyme dans la régénération de *Planaria subtentaculata* Drap. *Bull. Biol.* 203–27.

CHANDEBOIS, ROSINE (1965) Cell transformation systems in planarians. In *Regeneration in Animals*, Amsterdam, pp. 131–42.

CHICHKOFF, G.D. (1892) Recherches sur les Dendrocæles d'eau douce. *Arch. Biol.* **12**, 455–568.

CHILD, C.M. (1903a) Studies on regulation. I. Fission and regulation in *Stenostoma*. *Arch. Entwm.* **15**, I, II, pp. 187–237; III, pp. 355–420.

CHILD, C.M. (1903b) Experimental control of form-regulation in zooids and pieces of *Stenostoma*. *Arch. Entwm.* **15**, 603–37.

CHILD, C.M. (1904a) Studies on regulation. III. Regulative destruction of zooids and parts of zooids in *Stenostoma*. *Arch. Entwm.* **17**, 1–40.

CHILD, C.M. (1904b) Studies on regulation. IV. Some experimental modifications of form-regulation in *Leptoplana*. *J. Exp. Zool.* **1**, 95–134.

CHILD, C.M. (1904c). Studies on regulation. V. The relation between the central nervous system and regeneration in *Leptoplana*: posterior regeneration. *J.Exp.Zool.* **1**, 463–512.

CHILD, C.M. (1904d) Studies on regulation. VI. The relation between the central nervous system and regulation in *Leptoplana*: anterior and lateral regeneration. *J. Exp. Zool.* **1**, 513–58.

CHILD, C.M. (1905a) Studies on regulation. VII. Further experiments on form-regulation in *Leptoplana*. *J. Exp. Zool.* **2**, 253–85.

CHILD, C.M. (1905b) Studies on regulation. VIII. Functional regulation and regulation in *Cestoplana*. *Arch. Entwm.* **19**.

CHILD, C.M. (1905c) Studies on regulation. IX. The positions and proportions of parts during regulation in *Cestoplana* in the presence of cephalic ganglia. *Arch. Entwm.* **20**, 48–75.

CHILD, C.M. (1905d) Studies on regulation. X. The positions and proportions of parts during regulation in *Cestoplana* in the absence of the cephalic ganglia. *Arch. Entwm.* 157–86.

CHILD, C.M. (1906a) Contributions towards a theory of regulation. 1. The significance of the different methods of regulation in Turbellaria. *Arch. Entwm.* **20**.

CHILD, C. M. (1906b) The relation between regulation and fission in *Planaria*. *Biol. Bull.* **11,** 113–23.

CHILD, C. M. (1906c) Some considerations regarding so-called formative substances. *Biol. Bull.* **11,** 165–81.

CHILD, C. M. (1907) Some corrections and criticisms. *Arch. Entwm.* **24,** 131–46.

CHILD, C. M. (1909) The regulatory change of shape in *Planaria dorotocephala. Biol. Bull.* **16,** 277–96.

CHILD, C. M. (1910a) Analysis of form regulation with the aid of anesthetics. *Biol. Bull.* **18.**

CHILD, C. M. (1910b) The central nervous system as a factor in the regeneration of polyclad Turbellaria. *Biol. Bull.* **19,** 333–8.

CHILD, C. M. (1910c) Physiological isolation of parts and fission in *Planaria. Arch. Entwm.* **30,** Teil II, 159–205.

CHILD, C. M. (1911a) Experimental control of morphogenesis in the regulation of *Planaria. Biol. Bull.* **20,** 309–31.

CHILD, C. M. (1911b) Studies on the dynamics of morphogenesis and inheritance in experimental reproduction. I. The axial gradient in *Planaria dorotocephala* as a limiting factor in regulation. *J. Exp. Zool.* **10,** 265–320.

CHILD, C. M. (1911c) Studies on the dynamics of morphogenesis. II. Physiological dominance of anterior over posterior regions in the regulation of *Planaria dorotocephala. J. Exp. Zool.* **11,** 221–80.

CHILD, C. M. (1911d) Studies on the dynamics of morphogenesis. III. The formation of new zooids in *Planaria* and other forms. *J. Exp. Zool.* **11,** 187–220.

CHILD, C. M. (1911e) A study of senescence and rejuvenescence based on experiments with *Planaria dorotocephala. Arch. Entwm.* **31,** 537–612.

CHILD, C. M. (1911f) Die physiologische Isolation von Teilen des Organismus. *Vortr. u. Aufs. Entwm.* **11.**

CHILD, C. M. (1912) Studies on the dynamics of morphogenesis. IV. Certain dynamic factors in the regulatory morphogenesis of *Planaria dorotocephala*, etc. *J. Exp. Zool.* **13,** 103–52.

CHILD, C. M. (1913a) The asexual cycle of *Planaria velata* in relation to senescence and rejuvenescence. *Biol. Bull.* **25,** 187–203.

CHILD, C. M. (1913b) Studies on the dynamics, etc. V. The relation between resistance to depressing agents and rate of metabolism in *Planaria dorotocephala* and its value as a method of investigation. *J. Exp. Zool.* **14,** 153–206.

CHILD, C. M. (1913c) Studies on the dynamics, etc. VI. The nature of the axial gradients in Planaria and their relation to antero-posterior dominance, polarity and symmetry. *Arch. Entwm.* **37,** 108–58.

CHILD, C. M. (1913d) Certain dynamic factors in experimental reproduction and their significance for the problems of reproduction and development. *Arch. Entwm.* **35,** 598–641.

CHILD, C. M. (1914a) Susceptibility gradients in animals. *Science 29.*

CHILD, C. M. (1914b) Starvation, rejuvenescence and acclimatization in *Planaria dorotocephala. Arch. Entwm.* **38,** 418.

CHILD, C. M. (1914c) Studies on the dynamics, etc. VI. Dynamic factors in head determination in *Planaria. J. Exp. Zool.* **17,** 61–79.

CHILD, C. M. (1914d) Asexual breeding and prevention of senescence in *Planaria velata. Biol. Bull.* **26,** 286–93.

CHILD, C. M. (1914e) Studies on the dynamics, etc. VII. The stimulation of pieces by section in *Planaria dorotocephala. J. Exp. Zool.* **16,** 413–42.

CHILD, C. M. (1915a) *Individuality in Organisms,* Chicago.

CHILD, C. M. (1915b) *Senescence and Rejuvenescence,* Chicago.

CHILD, C. M. (1916) Studies on the dynamics, etc. IX. The control of head form and head-frequency in *Planaria* by means of potassium cyanide. *J. Exp. Zool.* **21**, 101.

CHILD, C. M. (1919a) A comparative study of carbon dioxide production during starvation in *Planaria*. *Amer. J. Phys.* **48**, 231–57.

CHILD, C. M. (1919b) The effect of cyanides on carbon dioxide production and on susceptibility to lack of oxygen in *Planaria dorotocephala*. *Amer. J. Phys.* **47**, 372–96.

CHILD, C. M. (1919c) Susceptibility to lack of oxygen during starvation in *Planaria*. *Amer. J. Phys.* **49**, 403–19.

CHILD, C. M. (1920) Some considerations concerning the nature and origin of physiological gradients. *Biol. Bull.* **39**, 147–187.

CHILD, C. M. (1921) Studies on the dynamics, etc. XI. Physiological factors in the development of the planarian head. *J. Exp. Zool.* **33**, 409–33.

CHILD, C. M. (1924) *Physiological Foundations of Behavior*, New York.

CHILD, C. M. (1929a) Physiological dominance and physiological isolation in development and reconstruction. *Arch. Entwm.* **117**, 21–66.

CHILD, C. M. (1929b) The physiological gradients. *Protoplasma* **5**, 447–76.

CHILD, C. M. (1930) The susceptibility of *Planaria* to potassium cyanide in relation to hydrion concentration and to certain changes in salt content of the medium. *Phys. Zool.* **3**, 90–135.

CHILD, C. M. (1932) Experimental studies on a Japanese planarian. I. Fission and differential susceptibility. *Sc. Rep. Tohogu Imp. Univ.* **7**, 313–45.

CHILD, C. M. (1941) *Patterns and Problems of Development*, Chicago.

CHILD, C. M. (1946) Organizers in development and the organizer concept. *Phys. Zool.* **19**, 89–147.

CHILD, C. M. (1948a) Differential oxidation and reduction of indicators in reconstitution of *Hydra* and a planarian. *Phys. Zool.* **21**, 327–50.

CHILD, C. M. (1949) Differential intracellular oxidation and reduction of indicators in relation to fission. *Phys. Zool.* **22**, 89–116.

CHILD, C. M. and MCKIE, A. V. M. (1912) The central nervous system in teratomorphic forms of *Planaria dorotocephala*. *Biol. Bull.* **22**, 39–59.

CHILD, C. M. and WATANABE, Y. (1935) The head frequency gradient in *Euplanaria dorotocephala*. *Phys. Zool.* **8**, 1–40.

CHRANOWA, ANNA (1939) Wiederholte Regeneration bei Planarien. *Arch. Entwm.* **139**, 65–77.

CLARK, R. B. and EVANS, S. M. (1961) The effect of delayed brain extirpation and replacement on caudal regeneration in *Nereis diversicolor*. *J. Embr. Exp. Morph.* **9**, 97–105.

CLARK, R. B. and MCCALLION, D. J. (1959) Specific inhibition of differentiation in the frog embryo by cell-free homogenates of adult tissues. *Can. J. Zool.* **37**, 133–6.

CLEAVE, C. D. VAN (1929) An experimental study of fission and reconstruction in *Stenostomum*. *Phys. Zool.* **2**, 18–58.

CLEMENT, H. (1944) Les acides pentosenucléiques et la régénération. *Ann. Soc. Roy. Zool. Belg.* **75**, 25–33.

COLDWATER, K. B. (1930) Action of X-rays on glutathione content and oxygen consumption of normal and regenerating planarians. *Proc. Soc. Exp. Biol. Med.* **27**, 1031–3.

COLDWATER, K. B. (1933) The effect of sulphydryl compounds upon regenerative growth. *J. Exp. Zool.* **65**, 43–71.

COLOSI, G. (1922) Il comportamente della superficie di taglio nelle planarie mutilate. *Rev. Sc. Nat.* **13**.

COLOSI, G. (1923) Considerazioni analitiche sulla individualità e sui rapporti con la rigenerazione e la riproduzione. *Boll. Mus. Anat. Comp. R. Univ. Torino*, **38**.

COWARD, S.J., FLICKINGER, R.A., and GAREN, E. (1964) The effect of nutrient media upon head frequency in regenerating planaria. *Biol. Bull.* **126**, 345–53.

CURTIS, W.C. (1902) The life history, the normal fission and the reproductive organs of *Planaria maculata. Proc. Boston Soc. Nat. Hist.* 515–59.

CURTIS, W.C. (1928) Old problems and a new technique. *Science* **67**, 141.

CURTIS, W.C. (1936) Effects of X-rays and radium upon regeneration. In *Biol. Effects of Radiation*, vol. I, pp. 411–58.

CURTIS, W.C. and HICKMAN, J. (1925) Effects of X-rays and radium upon regeneration in planarians. *Anat. Rec.* **34**, 145–6.

CURTIS, W.C. and SCHULZE, L.M. (1924) Formative cells of planarians. *Anat. Rec.* **29**, 105.

CURTIS, W.C. and SCHULZE, L.M. (1934) Studies upon regeneration. I. The contrasting powers of regeneration in *Planaria* and *Procotyla. J. Morph.* **55**, 477–512.

DAHM, A.G. (1958) *Taxonomy and Ecology of Five Species Group in the Family Planariidae*, Nya Lithografen, Malmö, 241 pp.

DALYELL, J.G. (1814) *Observations on some Interesting Phenomena in Animal Physiology exhibited by Several Species of Planariae*, Edinburgh.

DARWIN, CH. (1844) Brief descriptions of several terrestrial planariae and of some remarkable marine species. *Ann. Mag. Nat. Hist.* **14**, 241.

DESCOTILS, HEERNU F., QUERTIER, J. and BRACHET, J. (1961) Quelques effects du β-mercaptoéthanol et du dithiodiglycol sur la régénération. *Devel. Bio.* **3**, 277–96.

DIMMITT, J. and MARSH, G. (1942) Electrical control of morphogenesis in regenerating *Dugesia tigrina*. II. Potential gradients vs. current density as control factors. *J. Cell. Comp. Phys.* **40**, 11–24.

DOYLE, W.L. and OMOTO, J.H. (1950) Ultra-microdetermination of nitrogen. *Anal. Chem.* **22**, 603.

DRAPERNAULD, J.P.R. (1800/1) *Tableau des mollusques terrestres et fluviatiles de la France*, Montpellier, pp. 100–2.

DRESDEN, JR.D. (1940) Pharynx regeneration in *Polycelis nigra. Acta Néerl. Morphol. Norm. et Path.* **3**, 140–50.

DRIESCH, H. (1908) *The Science and Philosophy of the Organisms*, London.

DUBOIS, FRANÇOISE (1946a) Rythme saisonnier de ponte chez les Planaires d'eau douce, dans leur milieu naturel et en élevage. *C.R. Soc. Biol.* **140**, 881–4.

DUBOIS, FRANÇOISE (1946b) Influence de l'inanition et de la nutrition sur la ponte des Planaires d'eau douce. *C.R. Soc. Biol.* **140**, 884–5.

DUBOIS, FRANÇOISE (1948a) Sur les conditions de la migration des cellules de régénération chez les Planaires d'eau douce. *C.R. Soc. Biol.* **142**, 533.

DUBOIS, FRANÇOISE (1948b) Démonstration de la migration des cellules de régénération des Planaires par la méthode des greffes et des irradiations combinées. *C. R. Acad. Sc.* **226**, 1316–17.

DUBOIS, FRANÇOISE (1948c) Sur une nouvelle méthode permettant de mettre en évidence la migration des cellules de régénération chez les Planaires. *C. R. Soc. Biol.* **142**, 599–700.

DUBOIS, FRANÇOISE (1948d) Sur la migration des cellules de régénération chez les Planaires. *13th Congr. Intern. Zool. Paris.*

DUBOIS, FRANÇOISE (1949) Contribution à l'étude de la régénération chez planaires dulcicoles. *Bull. Biol.* **83**, 213–83.

DUBOIS, FRANÇOISE (1950) Continuité de la migration des néoblastes dans la régénération de la planaire *Euplanaria lugubris. C.R. Soc. Biol.* **144**, 1545.

DUGÈS, A. (1828) Recherches sur l'organisation et les meurs des Planariées. *Ann. des Sc. Nat.* **15**, 139.

VAN DUYNE, J. (1896) Über Heteromorphose bei Planarien. *Pflüger's Arch.* **64**, 589–74.

EVERS, J. (1947) *Bioelectric Fields and Growth*, Univ. Texas Press, pp. 151–61.

FARADAY, M. (1833) On the planariae. *Edin. New Phil. J.* **14**, 183–9.

FEDECKA-BRUNER, BARBARA (1961) La régénération de l'appareil copulateur chez la planaire *Dugesia lugubris*. *Arch. d'Anat. Micr. Morph. Exp.* **50**, 221–32.

FEDECKA-BRUNER, BARBARA (1964) Radiodestruction des testicules suivie de régénération chez la planaire *Dugesia lugubris*. *C.R. Acad. Sc. Paris* **258**, 3353–6.

FEDECKA-BRUNER, BARBARA (1965) Régénération des testicules des planaires après destruction par les rayons X. In *Regeneration in Animals*, Amsterdam.

FJERDINGSTAD, E.J., NISSEN, T., and RØIGAARD-PEDERSEN, H.H. (1965) Effect of ribonucleic acid (RNA) extracted from the brain of trained animals on learning in rats. *Scand. J. Psychol.* **6**, 1–6.

FLEXNER, S. (1898) The regeneration of the nervous system of *Planaria torva* and the anatomy of double-headed forms. *J. Morph. Phys.* **14**, 337–46.

FLICKINGER, REED A. (1959) A gradient of protein synthesis in planaria and reversal of axial polarity of regenerates. *Growth* **23**, 251–71.

FLICKINGER, REED A. (1962) Sequential gene action, protein synthesis, and cellular differentiation. *Intern. Rev. Cytol.* **13**, 75–98.

FLICKINGER, REED A. (1963a) Control of cellular differentiation by regulation of protein synthesis. *Amer. Zool.* **3**, 209–21.

FLICKINGER, REED A. (1963b) The site of incorporation of ^{14}C-labelled CO_2 and glycine in the planarian *Dugesia dorotocephala*. *Exp. Cell Res.* **30**, 605–7.

FLICKINGER, REED A. (1963c) Actinomysin D effects in the frog embryos: evidence for sequential synthesis of DNA-dependent RNA. *Science* **141**, 1063–4.

FLICKINGER, R.A. and BLOUNT, W. (1957) The relation of natural and imposed electrical potentials and respiratory gradients to morphogenesis. *J. Cell. Comp. Phys.* **50**, 403–22.

FLICKINGER, R.A. and COWARD, S.J. (1962) The induction of cephalic differentiation in regenerating *Dugesia dorotocephala* in the presence of the normal head and in unwounded tails. *Devel. Biol.* **5**, 179–204.

FRAENKEL, G. (1953) Studies on the distribution of vitamin B_T (carnitine). *Biol. Bull.* **104**, 359–71.

FRAPS, M. (1930) Studies on respiration and glycolysis in planaria. I. Methods and certain basic factors in respiration. *Phys. Zool.* **3**, 242–70.

FREISLING, M. and REISINGER, E. (1958) Zur Genese and Physiologie von Restitutionskörpern aus Planarien-Gewebsbrei. *Arch. Entwm.* **150**, 581–606.

FULINSKI, B. (1922) Über das Restitutionsvermögen der Rhabdocoelen. *Arch. Entwm.* **51**, 575–86.

GABRIEL, A. (1963) Influence du β-mercaptoéthanol sur la régénération de *Dugesia gonocephala* (Turbellarié, Triclade). *C. R. Acad. Sc. Paris* **257**, 976–8.

GABRIEL, A. (1965) Effets du β-mercaptoéthanol et de l'actinomycine B sur la régénération de la planaire *Dugesia gonocephala* (Turbellarié, Triclade). In *Regeneration in Animals*, Amsterdam, pp. 149–59.

GAZSÓ, L.R. (1958) Contribution à l'étude de la régénération du pharynx du *Dendrocoelum lacteum* et de la *Dugesia lugubris*. *Acta Biol. Acad. Sc. Hung.* **8**, 263–72.

GAZSÓ, L.R., TÖRÖK, L.J., and RAPPAY, G. (1961) Contributions to the histochemistry of the nervous system of planarians. *Acta Biol. Acad. Sc. Hung.* **11**, 411–28.

GEBHARDT, H. (1926) Untersuchungen über die Determination bei Planarienregeneraten. *Arch. Entwm.* **107**, 685–725.

GELEI, J. (1912) Studies in the histology of *Dendrocoelum lacteum* (Ørsted). (In Hungarian) (see Wetzel, 1964).

GELEI, J. (1913) Über die Ovogenese von *Dendrocoelum lacteum*. *Arch. Zellf.* **11**, 51–150.

GHIRARDELLI, E. (1959) Gli acidi nucleici nella rigenerazione di dischetti isolati dal corpo di *Dugesia lugubris*. *Acta Embr. Morph. Exp.* **2**, 314–28.

GHIRARDELLI, E. (1965) Differentiation of the germ cell and regeneration of the gonads in planarians. In *Regeneration in Animals*, Amsterdam, pp. 177–84.

GHIRARDELLI, E. and TASSELLI (1965) Sul numero degli occhi rigenerati in *Euplanaria lugubris. Atti Ac. Sc. Inst. Bologna. Sc. Fis. Ser.* **11**, 3, 1–8.

GIANFERRARI, LUISA (1929) Raggi X, rigenerazione e mortalità nelle Planarie *(Polycelis nigra, Planaria polychroa). Boll. Soc. Ital. Biol. Sper.* **4**, 915–20.

GOETSCH, W. (1921) Regeneration und Transplantation bei Planarien. *Arch. Entwm.* **49**, 359–82.

GOETSCH, W. (1922) Regeneration und Transplantation bei Planarien. II. *Arch. Entwm.* **51**, 251–5.

GOETSCH, W. (1925a) Regeneration und Determination. *Biol. Zentr. Bl.* **45**, 641–68.

GOETSCH, W. (1925b) *Tierkonstruktionen*, München.

GOETSCH, W. (1926) "Organisatoren" bei regenerativen Prozessen. *Naturw.* **46**, 1011–16.

GOETSCH, W. (1929) Das Regenerationsmaterial und seine Beeinflussung. *Arch. Entwm.* **117**, 211–311.

GOETSCH, W. (1932) Die Regeneration der Landplanarien und die Theorie der "Relativen Determination". *Naturwiss.* **20**, 960–2.

GOETSCH, W. (1946a) Vitamin T, ein neuartiger Wirkstoff. *Österr. Zool. Z.* **1**, 49–57.

GOETSCH, W. (1946b) Der Einfluß von Vitamin T auf Körperform und Entwicklung. *Naturwiss.* **33**, 149–54.

GOETSCH, W. (1950) Untersuchungen über den Wirkstoff "T" (T-Vitamin). *Österr. Zool. Z.* **2**, 435–70.

GOLDFARB, A.J. (1909) Influence of the nervous system in regeneration. *J. Exp. Zool.* **7**, 643–722.

GOLDFARB, A.J. (1911) The central nervous system in its relation to the phenomenon of regeneration. *Arch. Entwm.* **32**, 616–34.

GOLDSMITH, E.D. (1932) Abnormal outgrowths induced in planarians by cautery. *Anat. Rec.* **54**, 55–56.

GOLDSMITH, E.D. (1934) Correlations in planarian regeneration. *Proc. Nat. Acad. Sc. U.S.A.* **20**, 555–8.

GOLDSMITH, E.D. (1939) Regenerative and accessory growth in planarians. I. Inhibition of the development of regenerative and accessory growth. *Growth* **3**, 109–30.

GOLDSMITH, E.D. (1940) Regenerative and accessory growth in planarians. II. Initiation of the development of regeneration and accessory growth. *Phys. Zool.* **13**, 43.

GOLDSMITH, E.D. (1941) Further observations of supernumerary structures in individuals of an artificially produced clone of *Dugesia tigrina. Anat. Rec.* **81**, Suppl. 108–9.

GOLDSMITH, E.D. (1946) Thiourea and head regeneration in planarians. *Anat. Rec.* **96**, 85.

GOLDSMITH, E.D. and BLACK, H.M. (1948) The effect of podophyllin on regeneration in planaria. *Anat. Rec.* **101**, 708–9.

GRAFF, L. VON (1899) Monographie der Turbellarien. II. Tricladida Terricola (1912–17). Tricladida in *Bronn: Klassen und Ordnungen des Tierreichs*, 4, Abt. 1c, Pt. 2.

GRAFF, L. VON (1912–17) Turbellaria. II. *Bonn: Klassen und Ordnungen des Tierreiches.*

HALLEZ, P. (1892) Sur l'origine vraisemblablement tératologique de deux espèces de Triclades. *C.R.Acad. Sc. Paris.*

HALLEZ, P. (1899) Régénération et hétéromorphose. *Rev. Sc. Paris* **12**, 506–7.

HALLEZ, P. (1899–1900) Régénération comparée chez les Polyclades et les Triclades. *C. R. Ass. Fr.* 28e session, 270.

HAMBURGER, V. (1961) Regeneration. In *Encycl. Britt.*

HARTMANN, M. (1922) Über den dauernden Ersatz der ungeschlechtlichen Fortpflanzung durch fortgesetzte Regenerationen. *Biol. Centralbl.* **42**, 364–81.

HARVEY, W.H. (1857) *The Sea-side Book*, London.

HASLAUER, J. (1962) Der Einfluß des radonarmen bzw. radonfreien Gasteiner Thermalwassers auf die Regeneration von *Euplanaria gonocephala* (Dug). *Arch. Entwm.* 171–94.

HAUSER, P. J. (1956) Histological rebuilding processes in the planarian intestine during food absorption. *Mikroskopie* **11**, 20–31.

HAWK, P. B., OSER, R. L., and SUMMERSON, W. H. (1947) *Practical Physiological Chemistry*, N.Y., 820.

HEIN, CHARLOTTE (1928) Zur Kenntnis der Regenerationsvorgänge bei den Rhabdocoelen. *Z. W. Zool.* **130**, 469–546.

HENNEN, SALLY (1963) Chromosomal and embryological analyses of nuclear changes occurring in embryos derived from transfers of nuclei between *Rana pipiens* and *Rana sylvatica*. *Devel. Biol.* **6**, 133–82.

HESS, OLGA T. (1930) The effects of pure solutions of sodium, potassium, and calcium consumption of *Planaria dorotocephala*. *Phys. Zool.* **3**, 9–47.

HESSE, R. (1897) Die Augen der Plathelminthen insonderheit der tricladen Turbellarien. *Z. W. Zool.* **62**, 527–82.

HINRICHS, MARIE A. (1924a) A study of the physiological effects of caffein upon *Planaria dorotocephala*. *Biol. Bull.* **40**, 271–300.

HINRICHS, MARIE A. (1924b) A demonstration of the axial gradient by means of photolysis. *J. Exp. Zool.* **41**, 21.

HINRICHS, MARIE A. (1926) Modification of development on the basis of differential susceptibility to radiation. *Biol. Bull.* **50**.

HOFF-JØRGENSEN, E. and LØVTRUP, EBBA and SØREN (1953) Changes in deoxyribonucleic acid and total nitrogen in planarian worms during starvation. *J. Embr. Exp. Morph.* **1**, 161–5.

HOLMES, S. J. (1910/11) Minimal size reduction in planarians through successive regeneration. *J. Morph.* **22**, 989–92.

HULL, F. M. (1938) Regeneration and regulation of multiple heads in *Planaria maculata* and *Planaria agilis*. *Anat. Rec.* **72**, suppl. 86.

HUXLEY, J. S. and DE BEER, G. R. (1934) *The Elements of Experimental Embryology*, Cambridge.

HYMAN, LIBBIE H. (1919a) Physiological studies on planaria. I. Oxygen consumption in relation to feeding and starvation. *Amer. J. Phys.* **49**, 377–402.

HYMAN, LIBBIE H. (1919b) Physiological studies on planaria. II. *Amer. J. Phys.* **50**, 67.

HYMAN, LIBBIE H. (1919c) On the action of certain substances on oxygen consumption. II. Action of potassium cyanide on planaria. *Amer. J. Phys.* **48**, 340–71.

HYMAN, LIBBIE H. (1919d) Physiol. studies on planaria. III. *Biol. Bull.* **37**, 388–.

HYMAN, LIBBIE H. (1920) Physiological studies on planaria. IV. A further study of oxygen consumption during starvation. *Amer. J. Phys.* **53**.

HYMAN, LIBBIE H. (1923) Physiological studies on planaria. V. Oxygen consumption of pieces with regard to length, level and time after section. *J. Exper. Zool.* **37**, 47–68.

HYMAN, LIBBIE H. (1925) On the action of certain substances on oxygen consumption. VI. The action of acids. *Biol. Bull.* **49**, 288–.

HYMAN, LIBBIE H. (1928) Miscellaneous observations on *Hydra*, with special reference to reproduction. *Biol. Bull.* **54**, 65–97.

HYMAN, LIBBIE H. (1929) The effect of oxygen tension on oxygen consumption in planaria and some echinoderms. *Phys. Zool.* **2**, 505–34.

HYMAN, LIBBIE H. (1932a) Studies on the correlation between metabolic gradients, electrical gradients, and galvanotaxis. II. Galvanotaxis of the brown *Hydra* and some non-fissioning planarians. *Phys. Zool.* **5**, 185–.

HYMAN, LIBBIE H. (1932b) The axial respiratory gradient: experimental and critical. *Phys. Zool.* **5**, 566–92.

HYMAN, LIBBIE H. (1951) *The Invertebrates*, Vol. II.

HYMAN, L. H., WILLIER, B. H., and RIFENBURGH, S. A. (1924) Physiological studies on planaria. VI. A respiratory and histochemical investigation of the source of increased metabolism after feeding. *J. Exp. Zool.* **40**, 473–.

ICHIKAWA, A. and ISHII, SABURO (1961) Morphological and histochemical studies on regeneration in a freshwater planarian, *Dendrocoelopsis lacteus*. *J. Fac. Sc. Hokkaido Univ. Ser.* **6**, *Zool.* **14**, 595–606.

ICHIKAWA, A. and KAWAKATSU, M. (1964) A new freshwater planarian, *Dugesia japonica*, commonly but erroneously known as *Dugesia gonocephala* (Dugès). *Annot. Zool. Jap.* **37**, 185–94.

INOUE, K. (1961) Tissue culture analysis of embryonic cell differentiation in the presence of antibodies. *Acta Embr. Morph. Exp.* **4**, 183–208.

ISELY, M.F. (1925) The power of regeneration in three genera of planarians: *Planaria, Phagocata, Dendrocoelum. Anat. Rec.* **31**, 305–6.

ISHII, S. (1962/3) Electron microscopic observations on the planarian tissues. I. A survey of the pharynx. *Fukushima J. Med. Sc.* **9–10**, 51–73.

ISHII, S. (1964) The ultrastructure of the outer epithelium of the planarian pharynx. *Fukushima J. Med. Sc.* **11**, 109–25.

ISHII, S. (1965) Electron microscopic observations on the planarian tissues. II. The intestine. *Fukushima J. Med. Sc.* **12**, 67–87.

JAENICHEN, E. (1896) Beiträge zur Kenntnis des Turbellarien-Auges. *Z. W. Zool.* **62**.

JENKINS, MARIE M. (1959) The effects of thiourea and some related compounds on regeneration in planarians. *Biol. Bull.* **116**, 106–14.

JENKINS, MARIE M. (1961a) Respiration rates in planarians. II. The effect of goitrogens on oxygen consumption. *Biol. Bull.* **120**, 370–4.

JENKINS, MARIE M. (1961b) Respiration rates in planarians. III. The effect of thyroid compounds on oxygen consumption. *Biol. Bull.* **121**, 188–92.

JENKINS, MARIE M. (1963) Bipolar planarians in a stock culture. *Science* **142**, 1187.

JENNINGS, J.B. (1962) Further studies on feeding and digestion in triclad Turbellaria. *Biol. Bull.* **123**, 571–81.

JOHNSON, J.R. (1822) Observations on the genus Planaria. *Phil. Trans. Roy. Soc. Lond.*, Part II, 437–47.

JOHNSON, J.R. (1825) Further observations on Planariae. *Phil. Trans. Roy. Soc. Lond.*, Part II, 247–56.

KAESTNER, A. (1954/5) *Lehrbuch der speziellen Zoologie*, Teil I: *Wirbellose*. 1.Halbband.

KAHL, W. (1935) Regenerations- und Regulationsvorgänge während der Entwicklung überzähliger seitlicher Bildungen des Körperstammes von *Planaria gonocephala*. *Z. Wiss. Zool.* **146**, 621–77.

KANATANI, HARUO (1957a) Effects of crowding on the supplementary eye-formation and fission in the planarian, *Dugesia gonocephala*. *Annot. Zool. Jap.* **30**, 133–7.

KANATANI, HARUO (1957b) Studies on fission in the planarian, *Dugesia gonocephala*. I. Effects of heparin on the occurrence of fission. *J. Fac. Sc. Univ. Tokyo*, Sect. 4, 8, 17–21.

KANATANI, HARUO (1957c) Further studies on the effect of crowding on supplementary eye-formation and fission in the planarian, *Dugesia gonocephala*. *J. Fac. Sc. Univ. Tokyo*, Sect. 4.8, 23–39.

KANATANI, HARUO (1958a) Period of competence for supplementary eye-formation in the regenerating head of the planarian, *Dugesia gonocephala*. *J. Fac. Sc. Univ. Tokyo*, Sect. 4.8, part 2, 239–44.

KANATANI, HARUO (1958b) Formation of bipolar heads induced by demecolcine in the planarian, *Dugesia gonocephala*. *J. Fac. Sc. Univ. Tokyo*, Sect. 4.8, part 2, 254–70.

KANATANI, HARUO (1958c) Effect of demecolcine on head regeneration in the planarian, *Dugesia gonocephala*. *J. Fac. Sc. Univ. Tokyo*, Sect. 4, 8, 272.

KANATANI, HARUO (1958d) Effect of environment on the occurrence of supplementary eyes induced by lithium in the planarian, *Dugesia gonocephala*. *J. Fac. Sc. Univ. Tokyo*, Sect. 4.8, part 2.

KANATANI, HARUO (1959a) Effect of estradiol on head regeneration in the planarian, *Dugesia gonocephala*. *J. Fac. Sc. Univ. Tokyo*, Sect. 4. 8, part 3, 439–47.

KANATANI, HARUO (1959b) Studies on the formation of bipolar heads in the planarian, *Dugesia gonocephala*. I. Suppression of bipolar head formation by mono- and disaccharides, 2,4-dinitrophenol and sodium azide in transverse pieces and its relation to respiration of whole animals. *J. Fac. Sc. Univ. Tokyo*, Sect. 4, part 3, 449–66.

KANATANI, HARUO (1959c) Protective action of glucose and sucrose against retardation of head regeneration caused by demecolcine in the planarian, *Dugesia gonocephala*. *J. Fac. Sc. Univ. Tokyo*, Sect. 8, part 3, 467–71.

KANATANI, HARUO (1960a) Studies on the formation of bipolar heads in the planaria, *Dugesia gonocephala*. II. Suppression of bipolar head formation by potassium cyanide and anaerobiosis. *J. Fac. Sc. Univ. Tokyo* 49–58.

KANATANI, HARUO (1960b) Studies on fission in the planarian, *Dugesia gonocephala*. II. Effects of colchicine and demecolcine on the occurrence of fission. *J. Fac. Sc. Univ. Tokyo* 59–66.

KANATANI, HARUO (1962) Effects of nicotinamide on the growth and mitotic activity in regenerating planaria and bean root tips. *J. Fac. Sc. Univ. Tokyo* 387–95.

KAWAKATSU, M. (1959) An experimental study of the life-history of Japanese fresh-water planaria, *Phagocata vivida* (Ijima et Kaburati), with special reference to their fragmentation. *Bull. Kyoto Gakugei Univ.*, Ser. B, 35–59.

KEIL, EVA M. (1924) Studien über Regulationserscheinungen an *Polycelis nigra*. *Arch. Mikr. Anat. Entwm.* **102**, 452–88.

KEIL, EVA M. (1929) Regeneration in *Polychoerus caudatus*. *Biol. Bull.* **57**, 225.

KEILLER, V.H. (1911) A histological study of regeneration in short head pieces of *P. simplicissima*. *Arch. Entwm.* **31**, 131–44.

KELLER, J. (1894) Die ungeschlechtliche Fortpflanzung der Süßwasser-Turbellarien. *Jen. Zeits. f. Naturw.* **28**, 370–407.

KENK, R. (1924) Die Entwicklung und Regeneration des Kopulationsapparates der Planarien. *Zool. Jahrb.* **45**, 213-50.

KENK, R. (1930) Beiträge zum System der Protobursalier *(Tricladida paludicola)*, I–III. *Zool. Anz.* **89**.

KENK, R. (1932) A morphological proof of the existence of zooids in *Euplanaria dororocephala*. *Phys. Zool.* **8**, 442–56.

KENK, R. (1940) The reproduction of *Dugesia tigrina* (Girard). *Amer. Natur.* **74**, 471–5.

KENK, R. (1941) Induction of sexuality in *Dugesia tigrina* by transplantation. *Anat. Rec.* **79**, 12.

KEPNER, W.A. and CASH, J.R. (1915) Ciliated pits of *Stenostoma*. *J. Morph.* **26**, 234–46.

KIDO, T. (1952) Transplantation of planarian pieces into dorsal and ventral tissues. *Ann. Zool. Jap.* **25**, 383–7.

KIDO, T. (1957) Remarks on the so-called induction of the pharynx in planaria. *Sc. Rep. Kanazawa Univ.* 5.

KIDO, T. (1959) Location of the new pharynx in regenerating pieces of planaria. *Sc. Rep. Kanazawa Univ.* 6.

KIDO, T. (1961a) Studies on the pharynx regeneration in planarian, *Dugesia gonocephala*. I. Histological observation in translated pieces. *Sc. Rep. Kanazawa Univ.* 7.

KIDO, T. (1961b) Studies on the pharynx regeneration in planarian, *Dugesia gonocephala*. II. Histological observation in the abnormal regenerates produced experimentally. *Sc. Rep. Kanazawa Univ.* 7.

KIDO, T. and KISHIDA, YOSHIKAZU (1960) Effect of thiourea and iodine on the depigmentation of planarian eye. *Ann. Zool. Jap.* **33**.

KISHIDA, YOSHIKAZU (1961) Depigmentation of planarian eye treated with salts of dithiocarbamide and related chemicals. *Sc. Rep. Kanazawa Univ.* 7.

KLUG, H. (1960) Über die funktionelle Bedeutung der Feinstrukturen der exokrinen Drüsenzellen (Untersuchungen an Euplanaria). *Z. f. Zellforsch.* **51**, 617–32.

KOLMAYER, SIMONE and STÉPHAN-DUBOIS, F. (1960) Néoblastes et limitation du pouvoir de régénération céphalique chez la planaire *Dendrocoelum lacteum. J. Embr. Exp. Morph.* **8**, 376–86.

KONIECZNA, B., MARCZYŃSKA, B., and SKOWRON-CENDRZAK, A. (1960) Haematological and serological investigations in heteroparabiosis. *Fol. Biol.* **8**.

KONIECZNA, B., PLONKA, I., SKOWRON, A., and ZABINSKI, J. (1960) Haematological and serological investigations in heteroparabiosis after preimmunisation of one of the parabionts. *Fol. Biol.* **8**, 83–88.

KORSCHELT, E. (1927–9) *Regeneration and Transplantation*, 3 vols., Berlin.

KRATOCHWILL, K. W. (1962) Die Einwirkung von Röntgenstrahlen auf die Differenzierung in der Regeneration von *Euplanaria gonocephala. Z. Wiss. Zool.* **167**, 215–37.

LANG, A. (1884) *Die Polycladen. Fauna und Flora des Golfes von Neapel*, no. 11.

LANG, A. (1912) Über Regeneration bei Planarien. *Arch. Mikr. Anat.* **79**, 361–426.

LANG, A. (1913) Experimentelle und histologische Studien an Turbellarien, I–II. *Arch. Mikr. Anat.* **82**, 257.

LANGE, C. S. (1966) Observations on some tumours found in two species of planaria—*Dugesia etrusca* and *Dugesia ilvana. J. Embr. Exp. Morph.* **15**, 125–30.

LE MOIGNE, A. (1965a) Effets des irradiations aux rayons X sur le développement embryonaire et le pouvoir de régénération à l'éclosion de *Polycelis nigra* (Turbellarié, Triclade). *C. R. Acad. Sc. Paris* **260**, 4627–9.

LE MOIGNE, A. (1965b) Mise en évidence d'un pouvoir de régénération chez l'embryon de *Polycelis nigra. Bull. Soc. Zool. Fr.* **90**, 355–61.

LE MOIGNE, A. (1966) Etude du développement embryonaire et recherches sur les cellules de régénération chez l'embryon de la planaire *Polycelis nigra. J. Embr. Exp. Morph.* **15**, 39–60.

LE MOIGNE, A., SANZINE, M. J., LENDER, T., and DELAVAUET, R. (1965) Quelques aspects des ultrastructures du blastème de régénération et des tissues voisin chez *Dugesia gonocephala* (Turbellarié, Triclade). *C. R. Séanc. Soc. Biol.* **159**, 5–30.

LECAMP, M. (1942) *C. R. Acad. Sc. Paris* **214**, 330–2.

LEHNERT, G. H. (1891) Beobachtungen an Landplanarien. *Arch. f. Naturgesch.* **1**, 306–50.

LEMON, C. C. (1900) Notes on the physiology of regeneration of parts in *Planaria maculata. Biol. Bull.* 193–204.

LENDER, TH. (1950) Démonstration du role organisateur du cerveau dans la régénération des yeux de la planaire *Polycelis nigra* par la méthode des greffes. *C. R. Soc. Biol.* **144**, 1407.

LENDER, TH. (1951a) Découverte d'une planaire américaine, *Dugesia (Euplanaria) tigrina. Girard. Ass. Philom. d'Alsace et de Lorraine* **9**, 51.

LENDER, TH. (1951b) Sur les propriétés et l'étendue du champ d'organisation du cerveau dans la régénération des yeux de la planaire *Polycelis nigra. C. R. Séanc. Soc. Biol.* **145**, 1211.

LENDER, TH. (1951c) Sur les capacités inductrices de l'organisateur des yeux dans la régénération de la planaire *Polycelis nigra* (Ehr.): action du cerveau en voie de dégénérescence et en greffes hétéroplastiques. *C. R. Séanc. Soc. Biol.* **145**, 1378.

LENDER, TH. (1952a) Sur la régénération des yeux de la planaire *Polycelis nigra* en présence de broyats de la région antérieure du corps. *Bull. Biol. Fr. Belg.* **86**, 140–215.

LENDER, TH. (1952b) Le role inducteur du cerveau dans la régénération des yeux d'une planaire d'eau douce. *Ann. Biol.* **28**, 191–8.

LENDER, TH. (1954) Sur l'activité inductrice de la région antérieure du corps dans la régénération des yeux de la planaire *Polycelis nigra*: activité de broyats frais ou traités à la chaleur. *C. R. Soc. Biol.* **148**, 1859–61.

LENDER, TH. (1955a) Sur quelques propriétés de l'organisine de la régénération des yeux de la planaire *Polycelis nigra. C.R. Acad. Sc. Paris* **240**, 1725–8.

LENDER, TH. (1955b) Mise en évidence et propriétés de l'organisine de la régénération des yeux chez la planaire *Polycelis nigra. Rev. Suiss. Zool.* **62**, 268–75.

LENDER, TH. (1955c) Sur l'inhibition de la régénération du cerveau de la planaire *Polycelis nigra. C.R. Séanc. Acad. Sc.* **241**, 1863–5.

LENDER, TH. (1956a) Recherches expérimentales sur la nature et les propriétés de l'inducteur de la régénération des yeux de la planaire *Polycelis nigra. J. Embr. Exp. Morph.* **4**, 196–216.

LENDER, TH. (1956b) L'inhibition de la régénération du cerveau des planaires *Polycelis nigra* (Ehr.) et *Dugesia lugubris* (O.Schm.) en présence de broyats de têtes ou de queues. *Bull. Soc. Zool. Fr.* **81**, 192.

LENDER, TH. (1956c) Analyse des phénomènes d'induction et d'inhibition dans la régénération des planaires. *Ann. Biol.* **32**, 457–71.

LENDER, TH. (1960) L'inhibition spécifique de la différentiation du cerveau des planaires d'eau douce en régénération. *J. Embr. Exp. Morph.* **8**, 291–300.

LENDER, TH. (1963) Factors in morphogenesis of regenerating fresh-water planaria. In *Advances in Morphogenesis*, 2. Acad. Press Inc., New York, Ltd.

LENDER, TH. (1964) Mise en évidence et role de la neurosécrétion chez les planaires d'eau douce (Turbellariés, Triclades). *Ann. d'Endocrin.* **25**, 61–65.

LENDER, TH. (1965a) La régénération des Planaires. In *Regeneration of Animals*, Amsterdam, pp. 95–111.

LENDER, TH. (1965b) Quelques aspects de la régénération des planaires d'eau douce. In *Regeneration in Animals*, Amsterdam, pp. 143–8.

LENDER, TH. and DEUTSCH, V. (1961) L'activité inductrice du broyat d'embryon de Poulet au cours de la régénération des yeux de la Planaire *Polycelis nigra. C.R. Acad. Sc.* **253**, 550–1.

LENDER, TH. and GABRIEL, A. (1960) Etude histochimique des néoblastes de *Dugesia lugubris* (Turbellarié, Triclade) avant et pendent la régénération. *Bull. Soc. Zool. Fr.* **85**, 100–10.

LENDER, TH. and GABRIEL, A. (1961) La comportement des néoblastes pendent la régénération de la planaire *Dugesia lugubris* (Turbellarié, Triclade). *Bull. Soc. Zool. Fr.* **86**, 67–72.

LENDER, TH. and GABRIEL, A. (1965) Les néoblastes marqués par l'uridine tritiées migrent et édifient le blasteme de régénération des Planaires d'eau douce. *C.R. Acad. Sc.* **260**, 4095–7.

LENDER, TH. and GRIPON, P. (1962) La régénération des yeux et du cerveaux de *Dugesia lugubris* en présence de deux troncs nerveaux inégaux. *Bull. Soc. Zool. Fr.* **87**, 387–95.

LENDER, TH. and KLEIN, NICOLE (1961) Mise en évidence de cellules sécrétrices dans le cerveau de la planaire *Polycelis nigra*. Variation de leur nombre au cours de la régénération postérieure. *C.R. Acad. Sc.* **253**, 331–3.

LEVETZOW, K.G. v. (1939) Die Regeneration der polycladen Turbellarien. *Arch. Entwm.* **139**, 780–818.

LI, Y. (1928) Regulative Erscheinungen bei der Planarienregeneration unter anomalen Bedingungen. *Arch. Entwm.* **114**, 226–71.

LILLIE, F.R. (1900) Some notes on regeneration and regulation in planarians. *Amer. Nat.* **34**, 173–7.

LILLIE, F.R. (1901) A comparison of the power of regeneration in three genera of planarians, viz. *Planaria, Phagocata* and *Dendrocoelum. Science* **13**.

LILLIE, F.R. (1901) Notes on regeneration and regulation in planarians. *Amer. J. Phys.* **6**, 129–41.

LILLIE, F.R. and POULTON, F.P. (1898/97) On the effects of temperature on the development of animals. *Zool. Bull.* **1**, 180–2.

LINDH, N.O. (1956) The metabolism of nucleic acids during regeneration in *Euplanaria polychroa*. *Ark. f. Zool.*, Ser. 2, 9, 421–50.

LINDH, N.O. (1957a) Histological aspects on regeneration in *Euplanaria polychroa*. *Ark. f. Zool.*, Ser. 2, 11, 89–103.

LINDH, N.O. (1957b) The nucleic acid composition and nucleotide content during regeneration in the flatworm *Euplanaria polychroa*. *Ark. f. Zool.*, Ser. 2, 11, 153–66.

LINDH, N.O. (1957c) The mitotic activity during the early regeneration in *Euplanaria polychroa*. *Ark. f. Zool.*, Ser. 2, 10, 497–509.

LINDH, N.O. (1957d) *The Nucleic Acid Metabolism in Regenerating Flatworms*, Lund.

LLOYD, D.J. (1914) The influence of osmotic pressure upon the regeneration of *Gunda Ulvae*. *Proc. Roy. Soc.* **88**.

LONG, C.A. and HAYS, H.A. (1959) Some effects of colchicine on transverse fission in the planarian, *Dugesia dorotocephala*. *Trans. Kansas Acad. Sc.* **62**, 257–61.

LUND, E.J. (1921) Oxygen concentration as a limiting factor in the respiratory metabolism of *Planaria agilis*. *Biol. Bull.* **41**.

LUND, E.J. (1925) Experimental control of organic polarity by the electric current. *J. Exp. Zool.* **41**, 155–90.

LUND, E.J. (1928) Relation between continuous bio-electric currents and cell respiration. II. *J. Exper. Zool.* **51**, 265–90.

LUND, E.J. et al. (1947) *Bioelectric Fields and Growth*, Univ. Texas Press.

LUS, J. (1924) Studies on regeneration and transplantation in Turbellaria. I. Some considerations on polarity and heteromorphosis in fresh water planarians. *Bull. Soc. Nat. Mosc. Sect. Biol.*, exp. 1.

LUS, J. (1926) Regenerationsversuche an marinen Tricladen. *Arch. Entwm.* **108**, 203–27.

LØVTRUP, EBBA (1953) Studies on planarian respiration. *J. Exp. Zool.* **124**, 427–34.

MACARTHUR, J.W. (1920) Changes in acid and alkali tolerance with age in Planaria. *Amer. J. Phys.* **54**.

MACARTHUR, J.W. (1921) Gradients of vital staining and susceptibility in planaria and other forms. *Amer. J. Phys.* **57**.

MCCONNELL, J.V. (1966) Comparative physiology. Learning in invertebrates. *Ann. Rev. Physiol.* **28**, 107–36.

MCCONNELL, J.V. (1966) (Ed.) A manual of psychological experimentation in planarians. (To be had through *The Worm Runners' Digest*.)

MCCONNELL, J.V., JACOBSEN, A.L., and KIMBLE, D.R. (1959) The effects of regeneration upon retention of a conditioned response in the planarian. *J. Comp. Phys. Psychol.* **52**, 1–5.

MCWHINNIE, MARY A. (1955) The effect of colchicine on reconstitutional development in *Dugesia dorotocephala*. *Biol. Bull.* **108**, 54–65.

MCWHINNIE, MARY A. and GLEASON, MARY M. (1957) Histological changes in regenerating pieces of *Dugesia dorotocephala* treated with colchicine. *Biol. Bull.* **112**, 371–6.

MARINO, E. (1957) Experimental modifications of the rate of regeneration of the eye in *Dugesia lugubris*. *Ann. Med. Perugia* **48**, 359–69.

MARKERT, C.L. and URSPRUNG, HEINRICH (1963) Production of replicable persistent changes in zygote chromosomes of *Rana pipiens* by injected proteins from adult liver nucleus. *Devel. Biol.* **7**, 560–77.

MARSH, G. and BEAMS, H.W. (1952) Electrical control of morphogenesis in regenerating *Dugesia tigrina*. *J. Cell. Comp. Phys.* **39**, 191–213.

MEIXNER, M.J. (1938) Turbellaria (Strudelwürmer). In: GRIMPE and WAGLER: *Die Tierwelt der Nord- und Ostsee*, Teil IV B, Lief. 35.

MELANDER, Y. (1963) Cytogenetic aspects of embryogenesis in Paludicola, Tricladida. *Hereditas* **49**, 119–66.

MENGEBIER, W.L. and JENKINS, MARIE M. (1964) Succinoxidase activity in homogenates of *Dugesia dorotocephala*. *Biol. Bull.* **127**, 317–23.

MERKER, E. and GILBERT, H. (1932) Die Widerstandsfähigkeit von Süßwasserplanarien in ultraviolettreichem Licht. *Zool. Jahrb. Abt. Allgem. Zool. Phys.* **50**, 479.

MIFUNE, S. (1953) The effects of lithium chloride and low temperature on head regeneration of planaria. *Annot. Zool. Jap.* **26**, 32–37.

MILLER, C.A. and JOHNSON, W.H. (1959) Preliminary studies on the axenic cultivation of a planarian *(Dugesia)*. *Ann. N.Y. Acad. Sc.* **77**, 87–92.

MILLER, C.A., JOHNSON, W.H., and MILLIS, S.C. (1955) The sterilization and preliminary attempts in the axenic cultivation of the black planarian, *Dugesia dorotocephala*. *Indiana Acad. Sc.* **65**, 237–42.

MILLER, F.S. (1937) Some effects of strychnine in reconstitution in *Euplanaria dorotocephala*. *Phys. Zool.* **10**, 276–97.

MILLER, J.A. (1938) Studies on heteroplastic transplantation in Triclads. I. Cephalic grafts between *Euplanaria dorotocephala* and *E. tigrina*. *Phys. Zool.* **11**, 214–47.

LE MOIGNE, A. (1963) Etude du développement embryonaire de *Polycelis nigra*. *Bull. Soc. Zool. Fr.* **88**, 403–22.

MONTI, R. (1900) Rigenerazione nelle planarie marine. *Me. Inst. Lomb. Sc. e Lettr. Cl. Matem. Milano* **19**, 1–16.

MONTI, R. (1900) Studi sperimentali sulla rigenerazione nei *Rhabdoceli marini*. *Rend. Inst. Lombardo*, Ser. II, 33.

MORETTI, G. (1912) Sulla transposizione delle varie parte del corpo nella *Planaria torva*. *Arch. Ital. di Anat. e di Embr.* **10**, 437–60.

MORGAN, L.V. (1905) Incomplete anterior regeneration in the absence of the brain of *Leptoplana littoralis*. *Biol. Bull.* **9**, 187–93.

MORGAN, T.H. (1898) Experimental studies of the regeneration of *Planaria maculata*. *Arch. Entwm.* **7**, 364–97.

MORGAN, T.H. (1900a) Regeneration in planarians. *Arch. Entwm.* **10**, 58–119.

MORGAN, T.H. (1900b) Regeneration in *Bipalium*. *Arch. Entwm.* **9**, 563–86.

MORGAN, T.H. (1901) *Regeneration*, New York.

MORGAN, T.H. (1902a) Growth and regeneration in *Planaria lugubris*. *Arch. Entwm.* **13**, 179–212.

MORGAN, T.H. (1902b) The internal influences that determine the relative size of double structures in *Planaria lugubris*. *Biol. Bull.* **3**, 132–9.

MORGAN, T.H. (1904a) The control of heteromorphosis in *Planaria maculata*. *Arch. Entwm.* **17**, 693–95.

MORGAN, T.H. (1904b) Regeneration of heteromorphic tails in posterior pieces of *Planaria simplicissima*. *J. Exp. Zool.* **1**, 385–94.

MORGAN, T.H. (1904c) Notes on regeneration. (The limitation of the regenerative power of *Dendrocoelum lacteum*.) *Biol. Bull.* **6**, 159–72.

MORGAN, T.H. (1905) "Polarity" considered as a phenomenon of gradation of materials. *J. Exp. Zool.* **2**, 495–506.

MORGAN, T.H. and SCHIEDT, A.E. (1904) Regeneration in the planarian, *Phagocata gracilis*. *Biol. Bull.* **7**, 160–5.

MRÁZEK, A. (1914) Regenerationsversuche an der tripharyngealen *Planaria anophthalma*. *Arch. Entwm.* **38**, 252–76.

MURRAY, MARGARET R. (1928) The calcium–potassium ratio in culture media for *Planaria dorotocephala*. *Phys. Zool.* **1**, 137–46.

MURRAY, MARGARET R. (1931) *In vitro* studies of planarian parenchyma. *Arch. Exp. Zellf.* **11**, 656–68.

NAKAZAWA, S. (1961) The polarity theory of morphogenetic fields. *Sc. Rep. Tohoku Univ.* **4**, Ser. 57–92.

NEEDHAM, A. E. (1952) *Regeneration and Wound Healing*, London.

NEEDHAM, J. (1931) *Chemical Embryology*, Cambridge.

NISSEN, T., RØIGAARD-PEDERSEN, H. H., and FJERDINGSTAD, E. J. (1965) Effect of ribonucleic acid (RNA) extracted from the brain of trained animals on learning in rats. II. Dependence of RNA effect on training conditioned prior to RNA extraction. *Scand. J. Psychol.* **6**, 265–72.

NIU, M. C. (1963) The mode of action of ribonucleic acid. *Devel. Biol.* **7**, 379–93.

OGUKAWA, K. I. (1957) An experimental study of sexual induction in the asexual form of Japanese fresh-water planarian *Dugesia gonocephala* (Dugès). *Bull. Kyoto Gakugei Univ.*, Ser. B, **11**, 8–27.

OGUKAWA, K. I. and KAWAKATSU, M. (1956a) Studies on the fission of Japanese fresh-water planaria, *Dugesia gonocephala* (Dugès). IV. Comparative studies on the breeding and fission frequencies of sexual and assumed asexual races which were observed in laboratory cultures and natural habitats. *Bull. Kyoto Gakugei Univ.*, Ser. B, 23–42.

OGUKAWA, K. I. and KAWAKATSU, M. (1956b) Studies, etc. V. On the influence of fission frequencies of the animals of sexual and asexual races by means of head removal operations. *Bull. Kyoto Gakugei Univ.*, Ser. B, 43–59.

OGUKAWA, K. I. and KAWAKATSU, M. (1957) Studies, etc. VI. Comparative studies on breeding and fission frequencies of sexual and asexual races which were observed under by different food cultures. *Bull. Kyoto Gakugei Univ.*, Ser. B, 18–37.

OGUKAWA, K. I. and KAWAKATSU, M. (1958) Studies, etc. VII. Comparative studies on breeding and fission frequencies of sexual and asexual races influenced by various concentrations of Ringer solutions and hydrogen ion concentrations. *Ibid.* 19–44.

OKADA, Y. K. and KIDO, T. (1943) Further experiments on transplantation in planaria. *J. Fac. Sc. Tokyo Univ.* **4**, 601–23.

OKADA, Y. K. and SUGINO, H. (1934) Transplantation experiments in *Planaria gonocephala*, I–II. *Proc. Imp. Ac.* **10**, 37–40, 107–10.

OKADA, Y. K. and SUGINO, H. (1937) Transplantation experiments in *Planaria gonocephala* (Dugès). *Jap. J. Zool.* **7**, 373–439.

OLMSTED, J. M. D. (1918) The regeneration of triangular pieces of *Planaria maculata*. A study in polarity. *J. Exp. Zool.* **25**, 157–76.

OLMSTED, J. M. D. (1922) The role of the nervous system in the regeneration of polyclad Turbellaria. *J. Exp. Zool.* **36**, 48–56.

OOSAKI, T. and ISHII, S. (1965) Observations on the ultrastructure of nerve cells in the brain of the planarian *Dugesia gonocephala*. *Z. Zellf.* **66**, 782–93.

ORTNER, P. and SEILERN-ASPANG, F. (1962) Experimentelle Untersuchungen über die Aktivierung eines Organisationsfaktors bei Tricladen. *Z. Wiss. Zool.* **167**, 197–214.

OSBORNE, P. J. and MILLER, A. T., JR. (1962) Uptake and intracellular digestion of protein (peroxidase) in planarians. *Biol. Bull.* **123**, 589–96.

OSBORNE, P. J. and MILLER, A. T. (1964) Acid and alkaline phosphatase changes associated with feeding, starvation and regeneration in planarians. *Biol. Bull.* **125**, 285–92.

OTT, H. N. (1892) A study of *Stenostomum leucops* (O. Schm.). *J. Morph.* **263**, 304.

OWEN, E. E., WEISS, H. A., and PRINCE, L. H. (1938) Carcinogenetics and growth stimulation. *Science* **87**, 261–2.

OWEN, E. E., WEISS, H. A., and PRINCE, L. H. (1939) Carcinogens and planarian tissue regeneration. *Amer. J. Canc.* **35**, 424–6.

PALLAS, P. S. (1774) *Spicilegia zoologica quibus novae imprimis et obscurae animaliu. speciosiconibus atque conamentariis illustratur.* Fasc. X, Berolini.

PARKER, G. H. (1929) The metabolic gradient and its application. *Brit. J. Exp. Biol.* **6**, 412.

PASQUINI, P., GHIRARDELLI, E., and RUSTICALI, A. (1955) Sulla rigenerazione di sischetti isolati dal corpo delle planarie *(Planaria torva)*. *Atti Acad. Sc. Inst. Bologna. Cl. Sc. Fis.*, Ser. XI, 2.

PECHLANDER, R. (1957) Die Regenerationsfähigkeit von *Otomesostoma auditivum* (Forel et Duplessis) (Turbellaria). *Arch. Entwm.* **150**, 105–14.

PEDERSEN, K. J.(1956) On the oxygen consumption of *Planaria vitta* during starvation, the early phase of regeneration and asexual reproduction. *J. Exp. Zool.* **131**, 123–36.

PEDERSEN, K. J. (1958) Morphogenetic activities during planarian regeneration as influenced by triethylene melamine. *J. Embr. Exp. Morph.* **6**, 308–34.

PEDERSEN, K. J. (1959a) Some features of the fine structure and histochemistry of planarian subepidermal gland cells. *Z. Zellf.* **50**, 121–42.

PEDERSEN, K. J. (1959b) Cytological studies on the planarian neoblast. *Z. Zellf.* **50**, 799–817.

PEDERSEN, K. J. (1961a) Some observations on the fine structure of planarian protonephridia and gastrodermal phagocytes. *Z. Zellf.* **53**, 608–28.

PEDERSEN, K. J. (1961b) Studies on the nature of planarian connective tissue. *Z. Zellf.* **53**, 569–608.

PEDERSEN, K. J. (1961c) Studier over ferskvands tricladernes cytologi. *Kbh.* 1–94.

PEDERSEN, K. J. (1963) Slime-secreting cells in planarians. *Ann. N. Y. Acad. Sc.* **106**, 424–442.

PEDERSEN, K. J. (1964) The cellular organization of *Convoluta convoluta*. An acoel Turbellarian: A cytological histochemical and fine structural study. *Z. Zellf.* **64**, 655–87.

PEDERSEN, K. J. (1966) The organization of the connective tissue of *Discocelides Langi* (Turbellaria, Polycladidae). *Z. Zellf.* **81**, 94–117.

PENTZ, S. and SEILERN-ASPANG, F. (1961) Die Entstehung des Augenmusters bei *Polycelis nigra* durch Wechselwirkung zwischen dem Augenhemmungsfeld un der Augeninduktion durch das Gehirn. *Arch. Entwm.* **153**, 75–92.

PENZLIN, H. (1964) Die Bedeutung des Nervensystems für die Regeneration bei den Insekten. *Arch. Entwm.* **155**, 152–61.

PETTIBONE, M. and WULZEN, R. (1934) Variations in growth-promoting power for planarian worms of adult and embryonic tissues. *Phys. Zool.* **7**, 192–211.

POLEZEHAEV, L. V. (1947) Les recherches sur la régénération effectuées en U.R.S.S. de 1917 à 1947. (En russe.) *Sovietsk. Biol.* **24**, 267–8.

POPOFF, M. and PETTKOFF, P. (1924) Studien zur Beschleunigung der Wundregeneration durch Anwendung von Stimulationsmitteln. II. Versuche zur Beschleunigung der Wundgeneration an *Planaria gonocephala*. *Zellstimulationsforsch.* I, 57–73.

PRENANT, M. (1922) Recherches sur la parenchyme des Platyhelminthes. Essai d'histologie comparée. *Arch. Morph. Gén. Expér.* **5**, 1–175.

PRZIBRAM, H. (1921) Die Bruch-Dreifachbildung im Tierreich. *Arch. Entwm.* **48**.

PUCCINELLI, I. (1961) Variazioni del numero cromosomico e meccanismi di eliminazione cromosomica in poliploidi spermentali della planaria *Dugesia lugubris* (O. Schmidt). *Acta Embr. Morph. Exp.* **4**, 1–17.

RAND, H. W. (1924) Inhibiting agencies in ontogeny and regeneration. *Anat. Rec.* **29**, 98–99.

RAND, H. W. and BOYDEN, E. A. (1913) Inequality of the two eyes in regenerating planarians. *Zool. Jahrb.* **36**, 68–80.

RAND, H. W. and BROWN, A. (1926) Inhibition of regeneration in planarians by grafting: technique of grafting. *Proc. Nat. Acad. Sc.* **12**, 575–81.

RAND, H. W. and ELLIS, MILDRED (1926) Inhibition of regeneration in two-headed or two-tailed planarians. *Proc. Nat. Acad. Sc.* **12**, 570–4.

RANDOLPH, H. (1897) Observations and experiments on regeneration in planarians. *Arch. Entwm.* **5**.

RAVEN, C.P. and MIGHORST, J.C. (1948) On the influence of a posterior wound surface on the anterior regeneration in *Euplanaria lugubris* (Hesse). *Proc. Kön. Nederl. Ak. Wetensch.* **51,** 434–45.

REESE, D.H. (1964) An autoradiographic study of the origin of regenerated epidermis, acidophilic gland cells and pharynx in the planarian *Dugesia dorotocephala.* Thesis, Univ. Maryland, 28 pp.

Regeneration in Animals and Related Problems. Ed. V.KIORTSIS and H.A.L.TRAMPUSCH. North-Holland Publishing Co., Amsterdam, XXIV, 568, p. 1965.

REISS, P. (1935) Action biologique des rayons X et Y. *Cours de Physique Biol. de Vles.* 2, fasc. 2.

RITTER, W.E. and CONGDON, E.M. (1900) On the inhibition by artificial section of the normal fission plane in *Stenostoma. Proc. Calif. Acad. Sc.*, Ser. 3, **2,** 365–77.

ROBBINS, H.L. and CHILD, C.M. (1920) Carbon dioxide production in relation to regeneration in *Planaria dorotocephala. Biol. Bull.* **38,** 103–22.

ROBERTSON, J.A. (1927) Galvanotropic reactions of *Polycelis nigra* in relation to inherent electric polarity. *Brit. J. Exp. Biol.* **5,** 66.

ROBERTSON, J.A. (1930) On the rate of movement in a flatworm as influenced by body-level, duration of captivity and regeneration. *J. Exp. Biol.* **7,** 88–107.

RÖLICH, P. and TÖRÖK, L.J. (1961) Elektronenmikroskopische Untersuchungen des Auges von Planarien. *Z. Zellf.* **54,** 362–81.

ROSENBAUM, R.M. and ROLON, C.J. (1960) Intracellular digestion in the phagocytic cells of planaria. *Biol. Bull.* **118,** 315–23.

ROSENE, H. (1947) *Bioelectric Fields and Growth.* Univ. Texas Press, pp. 301–91.

RUHL, L. (1927a) Regenerationserscheinungen an Rhabdocoelen. *Zool. Anz.* **72,** 160–75.

RUHL, L. (1927b) Über Doppelbildungen und andere Mißbildungen bei *Stenostomum. Zool. Anz.* **72.**

RULON, O. (1936a) The effect of carbon dioxide, the hydrogen ion, calcium on the reconstitution in *Euplanaria dorotocephala. Phys. Zool.* **9,** 170–203.

RULON, O. (1936b) Experimental asymetries of the head of *Euplanaria dorotocephala. Phys. Zool.* **9,** 278–92.

RULON, O. (1937) The effects of certain organic acids on reconstitution in *Euplanaria dorotocephala. Phys. Zool.* **10,** 180–95.

RULON, O. (1938) Single and combined effects of cyanide and methylblue on reconstitution in *Euplanaria dorotocephala. Phys. Zool.* **11,** 202.

RULON, O. (1944) The control of reconstitutional development in *Dugesia dorotocephala* with ethyl alcohol. *Phys. Zool.* **17,** 152–8.

RULON, O. (1948) The control of reconstitutional development in planarians with sodium thiocyanate and lithium chloride. *Phys. Zool.* **21,** 231–6.

RULON, O. and CHILD, C.M. (1937) Experiments on scale of organization in *Euplanaria dorotocephala. Phys. Zool.* **10,** 396–404.

RUNNSTRÖM, J. (1928) Plasmabau und Determination bei dem Ei von *Paracentrotus lividus* Lm. *Arch. Entw. Mech.* **113.**

RUSTICA, C.P. (1924) The control of biaxial development in the reconstruction of pieces of planaria. *J. Exp. Zool.* **42,** 111.

SANTOS, F.V. (1929) Studies on transplantation in planaria. *Biol. Bull.* **57,** 188–97.

SANTOS, F.V. (1931) Studies on transplantation in planarians. *Phys. Zool.* **4,** 111–64.

SAXÉN, L. and WARTIOVAARA, J., HÄYRY, P., and VAINIO, T. (1965) Cell contact and tissue interaction in cytodifferentiation. *Rep. 4. Scand. Congr. Cell Res.* 21–36.

SAXÉN, L., TOIVONEN, S., and VAINIO, T. (1964) Initial stimulus and subsequent interactions in embryonic induction. *J.Embr. Exp. Morph.* **12,** 333–8.

SCHAPER, A. (1904) Experimentelle Untersuchungen über den Einfluß der Radiumstrahlen und der Radiumemanation auf embryonale und regenerative Entwicklungsvorgänge. *Anat. Anz.* **25**, 298–314, 326–37.

SCHAROV, J. (1934) Über die Abhängigkeit des Regenerationsprozesses vom Entwicklungsstadium und Alter bei *Dendrocoelum lacteum. Trav. Lab. Zool. Exp. et Morphol. Anim.* **3**, 141–64.

SCHEWTSCHENKO, N.N. (1936a) Polaritätsumkehr bei der Unterdrückung der dominierenden Region bei Planaria. *Zool. Anz.* **115**, 232–44.

SCHEWTSCHENKO, N.N. (1936b) Der Einfluß des Körperteiles von hoher physiologischer Aktivität auf den Regenerationsvorgang bei Planarien. *Proc. Zool. Biol. Inst. Charkov.*

SCHEWTSCHENKO, N.N. (1937) Die Wechselwirkung von Teilen von verschiedener physiologischer Aktivität bei Planarien. *Biol. Zentrbl.* **6**, 581–7.

SCHEWTSCHENKO, N.N. (1938) Migration of regeneration material in planaria. *Bull. Biol. Med. Exp. U.R.S.S.* **6**, 276–8.

SCHREIER, O. (1950) Die schädigende Wirkung verschiedener Chinone auf *Planaria gonocephala* Dug. und ihre Beziehung zur Childschen Gradiententheorie. *Österr. Zool. Z.* **2**, 70–116.

SCHULTZ, E. (1900) Über Regeneration bei Planaria. *Trav. Soc. Naturh. Pétersbourg*, 118–19.

SCHULTZ, E. (1901) Über Regeneration bei Polycladen. *Zool. Anz.* **24**, 527–9.

SCHULTZ, E. (1902a) Über das Verhältniss der Regeneration zur Embryonalentwicklung und Knospung. *Biol. Zentrbl.* **22**, 360–8.

SCHULTZ, E. (1902b) Aus dem Gebiet der Regeneration. II. Über die Regeneration bei Turbellarien. *Z. Wiss. Zool.* **72**, 130.

SCHULTZ, E. (1904) Über Reduktionen. I. Über Hungererscheinungen bei *Planaria lactea. Arch. Entwm.* **18**.

SCHULTZ, E. (1906) Über Reduktion. II. Über Hungererscheinungen bei *Hydra fusca. Arch. Entwm.* **21**.

SEILERN-ASPANG, F. (1957) Polyembryonie in der Entwicklung von *Planaria torva* (M. Schultz) auf Deckglaskultur. *Zool. Anz.* **159**, 193–202.

SEILERN-ASPANG, F. (1958) Entwicklungsgeschichtliche Studien an Paludicolen Tricladen. *Arch. Entwm.* **150**, 425–80.

SEILERN-ASPANG, F. (1960a) Beobachtungen über Zellwanderungen bei Trikladengewebe der Gewebekultur. *Arch. Entwm.* **152**, 35–42.

SEILERN-ASPANG, F. (1960b) Experimentelle Beiträge zur Frage der Zusammenhänge: Regenerationsfähigkeit-Geschwulstbildung. *Arch. Entwm.* **152**, 491–516.

SEILERN-ASPANG, F. (1960c) Syncytiale und differenzierte Tumoren bei Tricladen. *Arch. Entwm.* **152**, 517–23.

SEILERN-ASPANG, F. and KRATOCHWILL, K. (1965) Relation between regeneration and tumor growth. In *Regeneration in Animals*, pp. 452–73.

SEKERA, E. (1911) Weitere Beiträge zu den Doppelbildungen bei den Turbellarien. *Sitzber. Kgl. Böhm. Ges. Wiss. Prag.*

SENGEL, CATHERINE (1959) La région caudale d'une Planaire est-elle capable d'induire la régénération d'un pharynx? *J. Embr. Exp. Morph.* **7**, 73–85.

SENGEL, CATHERINE (1960) Culture *in vitro* de blastèmes de régénération de Planaires. *J. Embr. Exp. Morph.* **8**, 468–76.

SENGEL, CATHERINE (1963) Culture *in vitro* de blastèmes de régénération de la planaire *Dugesia lugubris. Ann. Epiphyties.* **14**, N. hors. sér. III, 173–83.

SENGEL, PH. (1951) Sur les conditions de la régénération normale du pharynx chez la planaire *Dugesia (= Euplanaria) lugubris* O. Schm. *C.R. Soc. Biol.* **145**, 1381.

SENGEL, PH. (1953) Sur l'induction d'une zone pharyngienne chez la Planaire d'eau douce *Dugesia lugubris* O. Schm. *Arch. Anat. Micr. Morph. Exp.* **42**, 57–66.

SENGEL, PH. (1964) Utilisation de la culture organotypique pour l'étude de la morpho-genèse et de l'endocrinologie chez les invertébrés. *Bull. Soc. Zool. Fr.* 10–41.

SHAW, G. (1791) Description of *Hirudo viridis*, a new English Leech. *Trans. Linn. Soc.* 1, 93–95.

SHEARER, C. (1930) A re-investigation of metabolic gradients. *J. Exp. Biol.* 7, 260–8.

SILBER, H. and HAMBURGER, V. (1939) The production of duplicitas cruciata and multiple heads by regeneration in *Euplanaria tigrina. Phys. Zool.* 12, 285–300.

SINGER, M. (1960) Nervous mechanisms in the regeneration of body parts in vertebrates. In: *18th Growth Sympos.*, Ronald Press Co., pp. 115–33.

SINGER, M. and MUTTERPERL, ELLEN (1963) Nerve fiber requirements for regeneration in forelimb transplants of the newt *Triturus. Devel. Biol.* 7, 180–91.

SIVICKIS, P.B. (1923) Studies on the physiology of reconstruction in *Planaria lata*, with a description of the species. *Biol. Bull.* 44, 113–52.

SIVICKIS, P.B. (1926) A convenient method for feeding planarians. *Science* 64.

SIVICKIS, P.B. (1930/31) A quantitative study of regeneration along the main axis of the triclad body. *Arch. Zool. Ital.* 16, 430–49.

SIVICKIS, P.B. (1931) A quantitative study of regeneration in *Dendrocoelum lacteum. Arb. d. II. Abt. Ungar. Biol. Forsch. Inst.* 4, 1.

SIVICKIS, P.B. (1933) Studies on the physiology of regeneration in Triclads. *Vytauto Didziojo Univ. Mat. Gamtos Fac. Darbai.* 369–441.

SIVICKIS, P.B. (1935) The effect of the regeneration activity upon rate of regeneration of eyes in *Planaria lugubris. V. D. Univ. Mat. Gamtos Fac. Darbai.* 9. *Zool.* 13–42.

SIVICKIS, P.B. (1935) Regeneration-reaction of an organism as a whole. *12. Congr. Intern. Zool. Lisboa*, C. R. 1.

SKAER, R.J. (1961) Some aspects of the cytology of *Polycelis nigra. Quart. J. Micr. Sc.* 5, 295–318.

SKAER, R.J. (1965) The origin and continuous replacement of epidermal cells in the planarian *Polycelis tenuis* (Iijima). *J. Embr. Exp. Morph.* 13, 129–39.

SMITH, AGENS A. and HAMMEN, C.S. (1963) Effects of metabolic inhibitors on planarian regeneration. *Biol. Bull.* 125, 534–41.

SONNEBORN, T.M. (1930) Genetic studies on *Stenostomum incaudatum* (nov. sp.). I. The nature and origin of differences among individuals formed during vegetative reproduction. *J. Exp. Zool.* 57, 57–108.

SPIRITO, A. (1935) Innesti e processi organizzativi in *Planaria torva. Arch. Zool. Ital.* 21, 37–65.

STEINBÖCK, O. (1954) Regeneration azöler Turbellarien. *Verh. D. Zool. Gesell. in Tübingen.* 86–93.

STEINBÖCK, O. (1954–6) Zur Theorie der Regeneration beim Menschen. *Forschungen und Forscher.* 4. (*Hyrtl-Almanach* I, 1958).

STEINBÖCK, O. (1963) Regenerations- und Konplantationsversuche an *Amphiscolop* spec. (*Turbellaria acoela*). *Arch. Entwm.* 154, 308–53.

STEINMANN, P. (1908) Untersuchungen über das Verhalten des Verdauungssystems bei der Regeneration der Tricladen. *Arch. Entwm.* 25, 523–68.

STEINMANN, P. (1910) Der Einfluß des Ganzen auf die Regeneration der Teile. Studien an Doppelplanarien. *Festschr. f. R. Hertwig.* 3, 29–53.

STEINMANN, P. (1916) *Regeneration in Bronn, Klassen und Ordnungen d. Tierr.* Vol 4. Vermes, Abt. Ic, Turbellarien, Lpz.

STEINMANN, P. (1925) Das Verhalten der Zellen und Gewebe im regenerierenden Tricladenkörper. *Verh. Naturf. Ges. Basel.* 36, 133–62.

STEINMANN, P. (1926) Prospective Analyse von Restitutionsvorgängen. *Arch. Entwm.* 108, 646–79.

STEINMANN, P. (1927) Prospektive Analyse von Restitutionsvorgänge. II. Über Reindividualisation d. i. Rückkehr von Mehrfachbildungen zur einheitlichen Organisation. *Arch. Entwm.* **112**, 333–49.

STEINMANN, P. (1928) Über Re-individualisation. *Rev. Suisse Zool.* **35**, 201–24.

STEINMANN, P. (1932) Über zellspezifischen Vitalfärbung als Mittel zur Analyse komplexer Gewebe. *Rev. Suisse Zool.* **39**, 397–410.

STEINMANN, P. (1933a) Vitalfärbungsstudien an explantierten Tricladengeweben und an ganzen Tieren. *Rev. Suisse Zool.* **40**, 265–8.

STEINMANN, P. (1933b) Transplantationsversuche mit vital gefärbten Tricladen. *Rev. Suisse Zool.* **40**, 269–72.

STEINMANN, P. (1933c) Vitale Färbungsstudien an Planarien. *Rev. Suisse Zool.* **40**, 529–58.

STEINMANN, P. and BRESSLAU, E. (1913) *Die Strudelwürmer (Turbellarien)*, Lpz. pp. 1–380.

STÉPHAN, F. (1962) Tumeurs spontanée chez la planaire *Dugesia tigrina*. *C. R. Soc. Biol.* **156**, 902–22.

STÉPHAN, F. (1963) Régénération des individues polypharyngés de *Dugesia tigrina* (planaires d'eau douce). *C.R. Soc. Biol.* **157**, 1063–5.

STÉPHAN, F. (1965) Régénération des formes doubles de *Dugesia tigrina* (triclades d'eau douce). In *Regeneration in Animals*, Amsterdam.

STÉPHAN-DUBOIS, F. (1951) Migrations et potentialités histogénétiques des cellules indifférenciées chez les hydres, les planaires et les oligochètes. *Ann. Biol.* **27**, 734–50.

STÉPHAN-DUBOIS, F. (1961) Les cellules de Régénération chez la planaire *Dendrocoelum lacteum. Bull. Soc. Zool. Fr.* **86**, 172–86.

STÉPHAN-DUBOIS, F. (1963) Régénération céphalique de l'Annellide *Lumbriculus variegatus*, à partir d'un territoire radiolésé. *C.R. Soc. Biol.* **157**, 1491–3.

STÉPHAN-DUBOIS, F. (1965) Les néoblastes dans la régénération chez les planaires. In *Regeneration in Animals*, Amsterdam, pp. 112–30.

STÉPHAN-DUBOIS, F. and GILGENKRANTZ, F. (1951a) Régénération après transplantation chez la planaire *Dendrocoelum lacteum* (O. F. M.). *C.R. Séanc. Soc. Biol.* **155**, 115–18.

STÉPHAN-DUBOIS, F. and GILGENKRANTZ, F. (1951b) Transplantation et régénération chez la planaire *Dendrocoelum lacteum. J. Embr. Exp. Morph.* **9**, 542–9.

STÉPHAN-DUBOIS, F. and KOLMAYER, S. (1959) La migration de la differentiation des cellules de régénération chez la planaire *Dendrocoelum lacteum. C.R. Soc. Biol.* **153**, 1856.

STÉPHAN-DUBOIS, F. and LENDER, TH. (1956) Corrélation humorales dans le régénération des planaires paludicoles. *Ann. Sc. Nat. Zool.* 11. ser.

STEVENS, N.M. (1901) Notes on regeneration in *Planaria lugubris. Arch. Entwm.* **13**, 396–409.

STEVENS, N.M. (1907) A histological study of regeneration in *Planaria simplicissima, P. maculata*, and *P.Morgani. Arch. Entwm.* **24**, 350–73.

STEVENS, N.M. (1909) Regeneration in *Planaria simplicissima* and *Planaria Morgani. Arch. Entwm.* **27**.

STEVENS, N.M. and BORIN, A.M. (1905) Regeneration in *Polychoerus caudatus. J. Exp. Zool.* **2**, 335–46.

STICH, H.F. (1959) Changes in nucleoli related to alterations in cellular metabolism. *Devel. Cytol.* Ed. RUDNICK, DOROTHEA, New York.

STOLTE, H.A. (1936) Die Herkunft des Zellmaterials bei regenerativen Vorgängen der wirbellosen Tiere. *Biol. Rev.* **11**, 1–48.

STOPPENBRINK, F. (1905a) Der Einfluß herabgesetzter Ernährung auf den histologischen Bau der Süßwassertricladen. *Z. Wiss. Zool.* **79**, 496–547.

STOPPENBRINK, F. (1905b) Die Geschlechtsorgane der Süßwassertricladen im normalen und im Hungerzustande. *Verh. Nat. Ver. Preuß. Rheinl.* **61**, 27–36.

STRANDSKOV, H.H. (1934) Certain physiological effects of X-rays on *Euplanaria dorotocephala*. I. Differences in susceptibility. *Phys. Zool.* **7**, 572–85.

STRANDSKOV, H.H. (1937) Certain physiological effects of X-rays on *Euplanaria doroto-cephala*. II. Regeneration. *Phys. Zool.* **10**, 14–20.

STRINGER, CAROLINE E. (1917) The means of locomotion in planarians. *Proc. Nat. Acad. Sc.* **3**, 691–2.

SUGINO, H. (1938) Miscellany on planaria transplantation. A supplemental note on the transplantation experiments in *Planaria gonocephala*. *Ann. Zool. Jap.* **17**, 185–93.

SUGINO, H. (1940) Influence of the head pieces implanted in the prepharyngeal region of planaria. *Ann. Zool. Jap.* **19**, 245–52.

SUGINO, H. (1941) Homopolar union in *Planaria gonocephala*. *Jap. J. Zool.* **9**, 175–83.

TARDENT, P. (1963) Regeneration in the Hydrozoa. *Biol. Rev.* **38**, 293–333.

TESHIROGI, W. (1955a) On the regeneration of a turbellarian, *Bdellocephala brunnea*, especially its frequencies of head, tail, pharynx, and genital organ formation. *Zool. Mag.* **64**.

TESHIROGI, W. (1955b) The effect of lithium chloride on head frequency in *Dugesia gonocephala* (Dugès). *Bull. Mar. Biol. St. Asamushi.* **8**.

TESHIROGI, W. (1956a) Effects of sodium thiocyanate and lithium chloride on regeneration in a turbellarian, *Bdellocephala brunnea*. *Zool. Mag.* **65**.

TESHIROGI, W. (1956b) Transplantation of the ganglionic region into the posterior levels of the turbellarian *Bdellocephala brunnea*. *Sc. Rep. Thoku Univ.* 22.

TESHIROGI, W. (1961) Transplantation of the ganglionic region of non-X-irradiated into the postpharyngeal region of X-irradiated ones, in *Dugesia gonocephala* (Dugès). *Sc. Rep. Hirosaki Univ.* **8**, 108–18.

TESHIROGI, W. (1962) Dynamic morphological change and time-tàble during the regeneration of the turbellarian *Bdellocephala brunnea*. *Sc. Rep. Fac. Lit. Sc. Hirosaki Univ.* **9**, 21–48.

TESHIROGI, W. (1963a) Fusion of different body segments, using X-ray-irradiated animals and non-irradiated ones in a freshwater planarian, *Bdellocephala brunnea*. *Sc. Rep. Fac. Lit. Sc. Hirosaki Univ.* **10**.

TESHIROGI, W. (1963b) Transplantation experiments of two short pieces of a freshwater planarian, *Bdellocephala brunnea*. *Jap. J. Zool.* **14**, 21–48.

TESHIROGI, W. and KUDO, S. (1962) Changes in the lipid content during regeneration of a planarian, *Bdellocephala brunnea*. *Zool. Mag. (Dobutsu Ugagaku Zasshi)* **71**, 196-201.

TESHIROGI, W. and MAIDA, J. (1956) On the depigmentation in the eye by thiourea treating and the skin-depigmentation by starvation in planarians. *Zool. Mag.* **65**, 377–81.

TESHIROGI, W. and NUMAKUNAI, T. (1962) Transplantation of the tail region into the anterior levels of a freshwater planarian, *Bdellocephala brunnea*. *Sc. Rep. Fac. Lit. Sc. Hirosaki Univ.* **9**, 86–94.

TESHIROGI, W. and OHBA, A. (1959) A consideration on the regenerative types based on the experiments in the Turbellaria, *Bdellocephala brunnea*. *Zool. Mag. (Dobutsugako Zasshi)* **68**, 307–16.

TESHIROGI, W. and YAMADA, M. (1960) On the frequencies of genital organ-formation in the transverse cut-pieces of a planarian, *Bdellocephala brunnea* cultured under various conditions. *Zool. Mag. (Dobutsugaku Zasshi)* **69**, 209–15.

THATCHER, H.F. (1902) The regeneration of the pharynx in *Planaria maculata*. *Amer. Natur.* **36**, 633–41.

TÖRÖK, L.J. (1958) Experimental contributions to the regenerative capacity of *Dugesia (= Euplanaria) lugubris*, O. Schm. *Acta Biol.* **9**, 79.

TÖRÖK, L.J. and RÖHRLICH, P. (1959) Contributions to the fine structure of the epidermis of *Dugesia lugubris* O. Schm. *Acta. Biol. Ac. Sc. Hung.* **10**. 23–48.

TOKIN, B. (1940) On the effect of carcinogenic substances on the process of regeneration. *C.T. Acad. Sc. U.R.S.S.* **29**, 519.

TOLMATSCHEWA-MELNITSCHENKO, E. P. (1939) Der Einfluß ultravioletter Strahlen auf den Regenerationsprozeß. I. Der Einfluß ultravioletter Strahlen auf die Regeneration von Planarien. *Bull. Biol. Med. Exp. U.R.S.S.* **7**, 361–5.

TRAMPUSCH, H. A. L. and HARREBOMÉE, A. E. (1965) Dedifferentiation a prerequisite of regeneration. In: *Regeneration in Animals*, Amsterdam, 341–76.

TURNER, R. (1946) Observations on the central nervous system of *Leptoplana acticola*. *J. Compar. Neurol.* **85**, 53–65.

UDE, J. (1908) Beiträge zur Anatomie und Histologie der Süßwassertricladen. *Z. Wiss. Zool.* **89**, 308–70.

URBANI, E. (1955) In *Experimentia* **209**.

URBANI, E. (1962) In *Advances in Morphogenesis* **2**, 61–68.

URBANI, E. (1964) Proteasi e morphogenesi. *Acta Vitaminol.* **18**, 103–24.

URBANI, E. (1965) Proteolytic enzymes in regeneration. In *Regeneration in Animals*, Amsterdam, pp. 39–55.

URBANI, E. and CECERE, F. (1964) Incorporazione di uridina H^3 nei neoblasti di planaria. *Rend. Inst. Sc. Univ. Camerino* **5**, 106–8.

URBANI, L., BELLINI, C., and ZAPPANICO, A. (1968) In *Ric. Scient.* **28**, 593.

VAN CLEAVE, C. D. (1934) The effects of X-radiation on the restitution of *Stenostomum tenuicauda* and some other worms. *Biol. Bull.* **47**, 304–14.

VANDEL, A. (1920) Sur la reproduction des Planaires et sur la signification de la fécondation chez ces animaux. *C. R. Acad. Sc. Paris* **171**, 125–8.

VANDEL, A. (1921a) Recherches expérimentales sur les modes de reproduction des planaires triclades paludicales. *Bull. Biol. Fr. Belg.* **55**, 343–518.

VANDEL, A. (1921b) La question de la spécificité cellulaire chez les Planaires. *C. R. Acad. Sc.* **172**, 1615–17.

VANDEL, A. (1921c) La régénération des glandes génitales chez les Planaires. *C. R. Acad. Sc.* **172**, 1072–4.

VANDEL, A. (1925) *Planaria subtentaculata* n'est q'une race asexuée de *Planaria gonocephala*. *Bull. Biol. Fr. Belg.* **59**, 498–505.

VANDEL, A. (1937) L'origine primordiale des cellules reproductrices. *Rev. Gén. Sc.*

VANNINI, E. (1965) Regeneration and sex gradient in some hermaphroditic animals. In *Regeneration in Animals*, Amsterdam, pp. 160–76.

VERHOEF, A. M. (1946) The mitotic activity during the regeneration of *Polycelis nigra*. *Proc. Kon. Ned. Ak. Wet.* **49**, 548–53.

VIAUD, G. (1954) Mise en évidence d'une force électro-motrice d'opposition produite par les planaires en réponse à une excitation électrique due à un courant continu. *Arch. Anat.* **37**, 145–51.

VIAUD, G. and MÉDIONI, J. (1949) Phototropism and regeneration in *Planaria lugubris*. O.Schm. *C. R. Séanc. Soc. Biol.* 1221–3.

VOIGT, W. (1894) Die ungeschlechtliche Fortpflanzung der Turbellarien. *Biol. Centrbl.* **14**, 745–51, 771–8.

VOIGT, W. (1899) Künstlich hervorgerufene Neubildung von Körperteilen bei Strudelwürmern. *Sitzber. Niederrh. Ges.*

WADDINGTON, C. H. (1956) *The Principles of Embryology*, Allen & Unwin, London.

WAGNER, F., VON (1890) Zur Kenntnis der ungeschlechtlichen Fortpflanzung von *Microstoma* nebst allgemeinen Bemerkungen über Teilung und Knospung im Tierreich. *Z. Jahrb.* **4**, 349–423.

WATANABE, Y. (1935a) Rate of head development as indicated by time of appearance of eyes in the reconstitution of *Euplanaria dorotocephala*. *Phys. Zool.* **8**, 41–64.

WATANABE, Y. (1935b) Head frequency in *Euplanaria maculata* in relation to the nervous system. *Phys. Zool.* **8**, 374–94.

WATANABE, Y. (1937a) Susceptibility of planarian pieces to various concentrations of Murray–Ringer solution. *Sc. Rep. Tohoku Univ. Biol.* **11**, 279.

WATANABE, Y. (1937b) Dominance and axial differentials in indophenol blue reaction during reconstitution in the stalkes medusa, *Haliclystus auriculata* Clark. *Sc. Rep. Tohoku Imp. Univ. Biol.* Ser. 4, **12**, 165–89.

WATANABE, Y. (1941a) Effects of modified Ringer solution on head reconstitution in *Dugesia (= Euplanaria) dorotocephala. Phys. Zool.* **14**, 316.

WATANABE, Y. (1941b) Experimental studies on a Japanese planarian. III. Effect of posterior section upon the rate of eye-formation in reconstitution. *Jap. J. Zool.* **9**, 533–44.

WATANABE, Y. (1941c) Experimental studies on a Japanese planarian. II. Axial differential in rate of eye formation in reconstitution. *Phys. Zool.* **14**, 437–48.

WATANABE, Y. (1948) Physiological studies on a fresh water triclad *Polycelis sapporo*. I. Character and rate of reconstitution of transverse pieces in relation to level of body. *J. Exp. Zool.* **109**, 291–330.

WATANABE, Y. and CHILD, C. M. (1933) Longitudinal gradient in *Stylochus ijimai*: with a critical discussion. *Phys. Zool.* **6**, 542.

WATANABE, Y. and CHILD, C. M. (1935) Differential reduction of methylene blue by *Corymorpha palma. Phys. Zool.* **8**, 395–416.

WATANABE, Y. and YAMAGISHI, S. (1955) Differential susceptibility and redox indicator patterns in *Polycelis sapporo. Phys. Zool.* **28**, 1–18.

WEIGAND, K. (1930) Regeneration bei Planarien und Clavilina unter dem Einfluß von Radiumstrahlen. *Z. Wiss. Zool.* **136**, 255–318.

WEIMER, B. R., PHILLIPS, C. W., and ANDERSON, D. M. (1938) The influence of thyroxin on eye formation in *Phagocata gracilis. Phys. Zool.* **11**, 158–67.

WERNER, O. *Reizphysiologische Untersuchungen an Planarien im U.V.-Licht.* Diss. Gießen.

WESERVE, F. G. and KENNEY, M. J. (1935) The biological effects of Roentgen rays on *Planaria dorotocephala. Amer. J. Roentg.* **33**, 386–9.

WETZEL, B. K. Studies on the fine structure of regenerating *Dugesia tigrina.* Diss. Harvard Univ. Cambridge, Mass.

WIERCENSKY, F. J. (1939) The effects on supersonic vibrations on head frequency in *Euplanaria dorotocephala. Phys. Zool.* **12**, 62–69.

WIGGLESWORTH, (1940) Local and general factors in the development of "pattern" in *Rhodinus prolixus* (Hemiptera). *J. Exper. Biol.* **17**, 180.

WILHELMI, J. (1909) Zur Regeneration und Polypharyngie der Tricladen. *Zool. Anz.* **34**, 673–7.

WILLIER, B. H., HYMNAN, L., and RIFENBURGH, S. A. (1925) A histochemical study of intracellular digestion in triclad flatworms. *J. Morph. Phys.* **40**, 299.

WILSON, J. W. (1925) The disintegration of *Planaria maculata* in KCN. *Proc. Amer. Soc. Zool. Anat. Rec.* **31**, 336.

WILSON, J. W. (1926–7) Regeneration of *Planaria maculata* in isotonic Ringer's fluid. *Anat. Rec.* **34**, 124.

WILSON, J. W. (1931) The disintegration of *Planaria maculata* in potassium cyanide in pond water and in diluted Ringer's fluid. *J. Exp. Zool.* **60**, 269–84.

WILSON, J. W. (1941) The regeneration of the planarian head in diluted Ringer's fluid. *J. Exper. Zool.* **86**, 225–46.

WOLFF, E. (1961) Migrations et contacts cellulaires dans la régénération. *Exper. Cell Res.* suppl. **8**, 246–59.

WOLFF, E. (1962) Recent researches on the regeneration of Planaria. In *Regeneration. 20th Growth Symposium,* The Ronald Press Co., pp. 53–84.

WOLFF, E. (1963) Self inhibitory factors in the regeneration of fresh water planarians. *Proc. 16th Internat. Congress Zool. Washington,* pp. 177–8.

WOLFF, E. and DUBOIS, F. (1947a) La migration des cellules de régénération et les facteurs qui la provoquent chez les Triclades. *C.R. Acad. Sc.* **224**, 1387–8.

WOLFF, E. and DUBOIS, F. (1947b) Sur une methode d'irradiation localisée permettant de mettre en évidence la migration des cellules de régénération chez les Planaires. *C.R. Soc. Biol.* **141**, 903–6.

WOLFF, E. and DUBOIS, F. (1947c) Sur les facteurs qui déclenchent la migration des cellules de régénération chez les Planaires. *C.R. Soc. Biol.* **141**, 906–9.

WOLFF, E. and DUBOIS, F. (1948a) Sur la migration des cellules de régénération chez les Planaires. *Rev. Suisse Zool.* **55**, 218–27.

WOLFF, E. and DUBOIS, F. (1948b) Mise en évidence d'un système de corrélations intervenant dans la régénération des Planaires d'eau douce. *Experientia* **4**, 273–5.

WOLFF, E. and HAFFEN, K. (1952) Sur une méthode de culture d'organes embryonaires *in vitro*. *Tex. Rep. Biol. Med.* **10**, 463–72.

WOLFF, E. and LENDER, TH. (1950a) Sur le déterminisme de la régénération des yeux chez une planaire d'eau douce *Polycelis nigra*. *C.R. Séance. Soc. Biol.* **144**, 1213.

WOLFF, E. and LENDER, TH. (1950b) Sur le role organisateur du cerveau dans la régénération des yeux chez une planaire d'eau douce. *C.R. Acad. Sc.* **230**, 2238–9.

WOLFF, E., SENGEL, P., and SENGEL, CATHERINE (1958) La région caudale d'une planaire est-elle capable d'induire la régénération d'une pharynx? *C.R. Séanc. Acad. Sc.* **246**, 1744–6.

WOLFF, E., LENDER, T., and ZILLER-SENGEL, CATHERINE (1964) Le role de facteurs auto-inhibiteurs dans la régénération des Planaires. *Rev. Suisse Zool.* **71**, 75–98.

WOLSKY, A.A. (1935) Starvation and regeneration potency in *Dendrocoelum lacteum*. *Nature* **135**, 102.

WOODRUFF, L. and BURNETT, A.L. (1965) The origin of blastemae cells in *Dugesia tigrina*. *Exper. Cell Res.* **38**, 295–305.

WULZEN, R. (1916) The pituitary gland. Its effect on growth and fission of planarian worms. *J. Biochem.* **25**, 625–33.

WULZEN, R. and BAHRS, A.M. (1935) A dietary factor which imparts to certain mammalian tissues a quality necessary to the correct nutrition of planarian worms. *Phys. Zool.* **8**, 457–73.

WULZEN, R. and BAHRS, A.M. (1936) A common factor in planarian and mammalian nutrition. *Phys. Zool.* **9**, 508–29.

WYMAN, J. (1865) An account of some experiments on planaria showing their power of repairing injuries. *Boston Soc. Nat. Hist.* **9**, 157.

YAMADA, T. and KARASAKI, S. (1963) Nuclear RNA synthesis in Newt iris cells engaged in regenerative transformation into lens cells. *Devel. Biol.* **7**, 595–604.

YAMAMOTO, T.S. (1957) Histochemistry of the fresh-water planarian *Dendrocoelopsis sp. Annot. Zool. Japon.* **30**, 150–5.

ZACHARIAS, O. (1885) Zur Frage der Fortpflanzung durch Querteilung bei Süßwasserplanarien. *Zool. Anz.* **8**.

ZACHARIAS, O. (1886) Über Fortpflanzung durch spontane Querteilung bei Süßwasserplanarien. *Z. Wiss. Zool.* **43**, 271–5.

ZELENY, C. (1917) The effect of degree of injury, level of cut and time within the regeneration cycle upon the rate of regeneration. *Proc. Nat. Acad. Sc. USA* **3**, 211–17.

ZILLER-ZENGEL, CATHERINE (1965) Inhibition de la régénération du pharynx chez les planaires. In *Regeneration in Animals*, Amsterdam, pp. 191–201.

ZWEIBAUM, J. (1915) La régénération des ovaires chez *Polycelis nigra*. *Arch. Entwm. Mech.* 430–71.

APPENDIX

Several papers have been sent to me after my book has been given to the printers. The publishers have kindly permitted me to mention them briefly in this appendix.

AKIN, GWYNN C. (1966) Self-inhibition of growth in *Rana pipiens* tadpoles. *Phys. Zool.* **39,** 341–56.

Inhibition of growth rate by species-specific substances has been shown to exist in crowded conditions, as has been shown for various other organisms. (It should be interesting to see if such substances are akin to inhibitors during regeneration.)

BETCHAKU, T. (1967) Isolation of planarian neoblasts and their behavior *in vitro* with some aspects of the mechanism of the formation of regeneration blastems. *J. Exp. Zool.* **164,** 407–34.

The author succeeded in isolating neoblasts and other parenchyma cells and gastro-dermal cells *in vitro*. This important new technique will, I am sure, in the future con-stitute a very valuable tool in studying physiological and biochemical alterations of the neoblasts during true regeneration. The author found that the mobility of the neo-blasts *in vitro* is poorly developed. (I think that this is quite understandable because under these conditions they lack the impetus to migrate, due to absence of "wound hormones".) The gastrodermal and "fixed" parenchymal cells migrate to the surface of the cell clusters *in vitro*. (The conditions in such clusters are, of course, quite differ-ent from those in the tissues near the wound in normal worms. I should therefore strongly question the author's hypothesis that the neoblasts do not migrate during normal regeneration. Such migrations over long distances have been proved to exist.)

BORIANI, G. and GHIRARDELLI, E. (1966) Azione dei raggi X sull' RNA dei neoblasti di *Polycelis nigra* in rigenerazione. *Radiobiol., Radioter. e Fis. Medica* **21,** 162–77.

A group of normal worms were irradiated with 1800 r, another with 2500 r, the latter showing a slower rate of regeneration, which confirms the results of other authors. Histophotometric measurements show that RNA content in the cytoplasm of the neoblasts is considerably diminished. The authors conclude that the slower regeneration rate is due to a reduced rate of protein synthesis due to diminished RNA activity.

BRANDI, LUISA and GHIRARDELLI, E. (1963) L'azione del cervello sulla rigenerazione delle gonadi di *Dugesia lugubris*. *Acc. Naz. Lincei. Classe Sc. Mat. Nat.* **35,** 120–5.

The head has seemingly an inductive effect on the maturation of the gonads.

GHIRARDELLI, E. and GORDINI, STEFANIA (1964) Influenza delle regione cefalica sulla rigenerazione delle gonadi in esemplari bicefali di *Dugesia lugubris*. *Acc. Naz. Lincei. Classe Sc. Fis. Mat. Nat.* **37,** 92–96.

The experiments in this paper indicate that some influence from the brain is necessary for the development of gonads. The authors decapitated the worms, and from the wound surface a median cut is made to the pharynx; in this way a double headed animal will eventually regenerate. After 10 days all gonads disappear. After 30 days each half regenerate gonads, but only on the lateral parts.

KISHIDA, Y. (1967) Electron microscope studies on the planarian eye. I. Fine structure of the normal eye. II. Fine structure of the regenerating eye. *Sc. Rep. Kanazawa Univ.* **12,** 95–109, 111–42.

These interesting papers give, for the first time, very detailed and instructive information on the regenerating eye. The experimental species is *Dugesia japonica*.

LE MOIGNE, A. (1966?) Premières recherches sur les cellules de régénération chez l'embryon de la Planaire *Polycelis nigra*. *Mém. Soc. Sc. Nat. et Mat. de Cherbourg* 73–89.

Embryos can regenerate before the nervous system is differentiated. After irradiation the neoblasts are incapable of promoting regeneration. Therefore, neoblasts exist as separate cells in the embryo.

LE MOIGNE, A. (1966) Etude au microscope électronique de cellules d'embryons de *Polycelis* (Turbellarié, Triclade), au début de leur développement. *C.R. Acad. Sc. Paris* **263**, 550–3.

The embryos contain embryonic cells which up until the formation of the transitory pharynx have distinct characteristics, namely absence of ergastoplasm, free ribosomes, poorly constructed mitochondria, but important exchanges take place between nucleus and cytoplasm.

LE MOIGNE, A., SANZIN, MARIE-JOSEPHE, and LENDER, T. (1966) Comparaison de l'ultrastructure du néoblaste et de la cellule embryonaire des Planaires d'eau douce. *C.R. Acad. Sc. Paris* **263**, 627–9.

The authors found that the neoblasts are embryonic cells which during development have attained a state of physiological rest.

ROSE, S. MERYL (1963) Polarized control of regional structure in *Tubularia*. *Devel. Biol.* **7**, 488–501.

By using bioelectric fields the author found that inhibitory substances are transmitted through the body of *Tubularia*. The substances can be moved electrophoretically as positively charged particles at pH 8.1. Perhaps the inhibitors are histones produced by nucleotide segments of DNA as proposed by Huang and Bonner (1962).

ROSE, S. MERYL (1965) Regeneration versus non-regeneration. *Bull. Tulane Univ. Med. Fac.* **24**, 237–43.

In this short but very stimulating paper covering the broad aspects of regeneration the author emphasizes the notion of the existence of genetic information in practically every cell of the body, and he suggests that the information moves only when the proper bioelectric condition is present.

ROSE, S. MERYL (1966) Polarized inhibitory control of regional differentiation during regeneration in *Tubularia*. II. Separation of active materials by electrophoresis. *Growth* **30**, 429–47.

This very interesting paper must be read in conjunction with the 1963 paper. The regional specificity is lost after extraction and general inhibition is lost at pH 4.1. Inhibition is removed by trypsin, pointing to the protein nature of the inhibiting substances.

ROSE, S. MERYL and POWERS, J.A. (1966) Polarized control of regional structure in *Tubularia*. I. The effect of extracts from distal and proximal regions. *Growth* **30**, 419–27.

Reconstituting proximal primordia of *Tubularia* were cultured in water with aqueous solutions of homogenates of adult hydranths, either distal or proximal. If distal solutions were used then the primordia did not regenerate distal parts; if proximal, then the primordia had their structures impaired. Extracts were made from primordia and added to the culture water with younger, less stable primordia, which were stabilized by distal extracts but inhibited in distal regeneration. Proximal structures were not stabilized by proximal extracts.

ROSE, S. MERYL and ROSE, FLORENCE C. (1965) The control of growth and reproduction in freshwater organisms by specific products. *Mitt. Intern. Verein Limnol.* **13**, 21–35.

In this paper the broader aspects of inhibition in crowded populations are discussed. (See also the paper by Akin.)

SAUZIN, MARIE-JOSEPHE (1966) Etude au microscope électronique du néoblaste de la Planaire *Dugesia gonocephala* (Turbellaria, Triclade) et ses changements ultra structuraux au cours des premiers stades de la régénération. *C.R. Acad. Sc. Paris* **263**, 605–8.

Before regeneration the neoblasts contain many free ribosomes but few other organites. During activation, when the blastema is being formed, the cytoplasm increases, the ribosomes are arranged in rosettes. The ergastoplasm, the Golgi apparatus and lipids appear, the number of mitochondria increases and they develop more structures.

STEINBÖCK, O. (1967) Regenerationsversuche mit *Hofstenia* Giselae Steinb. (Turbellaria acoela). *Arch. Entwm.* **158**, 394–458.

The author, who, even after the investigations by Pedersen, is not convinced that the parenchyme of the Acoeles is made of individual cells, holds that the regeneration of anterior organs is due to an a-cellular parenchyme. Regeneration cells "wurden niemals festgestellt". The paper is very interesting on account of the very varied experimental procedures and its theoretical discussions.

STEPHAN, F. and SCHILT, J. (1966) Etude histologique d'excroissance induites par cauterisation chez la planaire *Dugesia tigrina* Girard. *C.R. Acad. Sc. Paris* **263**, 1732–4.

After cauterization, both pre- and postpharyngeally, formations appear, which, if anterior, may produce eyes and in some cases nervous trunks. The parenchyme is dense and shows acidophilic hypersecretions.

TUCKER, MARIE (1959) Inhibitory control of regeneration in nemertean worms. *J. Morph.* **105**, 569–600.

In this interesting and informative paper the author confirms that in *Lineus vigetus* a hierarchy of differentiation patterns exists, each differentiated region inhibits the regeneration of its like in the succeeding region.

VANNINI, E. (1966) Alcune aspette attuali dei probleme della rigenerazione. *Arch. Zool. Ital.* **51**, 951–8.

Regeneration is controlled by the same genetic information as in embryonic and postembryonic ontogenesis. Only, the regeneration cells are not identical with the original embryonic cells. Their activity is regulated by inductive fields or gradients depending upon organizer centres and self-differentiation within the regeneration blastema.

WOLFF, E. (1966) General introduction. General factors of embryonic differentiation. In: *Cell Differentiation and Morphogenesis*. Intern. lecture course, Wageningen, The Netherlands, April 26–29, 1965. North-Holland Publ. Co., Amsterdam, pp. 1–23.

This thoughtful and stimulating paper should be read by all concerned with morphogenesis, and also by those dealing with planarian regeneration.

AUTHOR INDEX

SUBJECT AND SPECIES INDEX

Systematic names italicized; italicized page numbers indicate pages which include figures referring to species

271

OTHER TITLES IN THE ZOOLOGY DIVISION

General Editor: G. A. KERKUT